On the convergence of $\sum c_k f(n_k x)$

MEMOIRS
of the
American Mathematical Society

Number 943

On the convergence of $\sum c_k f(n_k x)$

István Berkes
Michel Weber

September 2009 • Volume 201 • Number 943 (second of 5 numbers) • ISSN 0065-9266

American Mathematical Society
Providence, Rhode Island

2000 *Mathematics Subject Classification.*
Primary 42C15, 42A55, 42A61, 30B50, 11K38, 60G50.

Library of Congress Cataloging-in-Publication Data

Berkes, Istvan, 1947–
On the convergence of $\sum c_k f(n_k x)$ / Istvan Berkes, Michel Weber.
p. cm. — (Memoirs of the American Mathematical Society, ISSN 0065-9266 ; no. 943)
"Volume 201, number 943 (second of 5 numbers)."
Includes bibliographical references and index.
ISBN 978-0-8218-4324-6 (alk. paper)
1. Convergence. 2. Fourier analysis. I. Weber, Michel, 1949– II. Title.

QA295.B47 2009
515′.24—dc22 2009019383

Memoirs of the American Mathematical Society

This journal is devoted entirely to research in pure and applied mathematics.

Subscription information. The 2009 subscription begins with volume 197 and consists of six mailings, each containing one or more numbers. Subscription prices for 2009 are US$709 list, US$567 institutional member. A late charge of 10% of the subscription price will be imposed on orders received from nonmembers after January 1 of the subscription year. Subscribers outside the United States and India must pay a postage surcharge of US$65; subscribers in India must pay a postage surcharge of US$95. Expedited delivery to destinations in North America US$57; elsewhere US$160. Each number may be ordered separately; *please specify number* when ordering an individual number. For prices and titles of recently released numbers, see the New Publications sections of the *Notices of the American Mathematical Society.*

Back number information. For back issues see the *AMS Catalog of Publications.*

Subscriptions and orders should be addressed to the American Mathematical Society, P. O. Box 845904, Boston, MA 02284-5904 USA. *All orders must be accompanied by payment.* Other correspondence should be addressed to 201 Charles Street, Providence, RI 02904-2294 USA.

Memoirs of the American Mathematical Society (ISSN 0065-9266) is published bimonthly (each volume consisting usually of more than one number) by the American Mathematical Society at 201 Charles Street, Providence, RI 02904-2294 USA. Periodicals postage paid at Providence, RI. Postmaster: Send address changes to Memoirs, American Mathematical Society, 201 Charles Street, Providence, RI 02904-2294 USA.

This publication is indexed in *Science Citation Index*®, *SciSearch*®, *Research Alert*®, *CompuMath Citation Index*®, *Current Contents*®*/Physical, Chemical & Earth Sciences.*
Printed in the United States of America.

∞ The paper used in this book is acid-free and falls within the guidelines established to ensure permanence and durability.
Visit the AMS home page at http://www.ams.org/

10 9 8 7 6 5 4 3 2 1 14 13 12 11 10 09

Contents

Abstract

Let f be a periodic measurable function and (n_k) an increasing sequence of positive integers. We study conditions under which the series $\sum_{k=1}^{\infty} c_k f(n_k x)$ converges in mean and for almost every x. There is a wide classical literature on this problem going back to the 30's, but the results for general f are much less complete than in the trigonometric case $f(x) = \sin x$. As it turns out, the convergence properties of $\sum_{k=1}^{\infty} c_k f(n_k x)$ in the general case are determined by a delicate interplay between the coefficient sequence (c_k), the analytic properties of f and the growth speed and number-theoretic properties of (n_k). In this paper we give a general study of this convergence problem, prove several new results and improve a number of old results in the field. We also study the case when the n_k are random and investigate the discrepancy the sequence $\{n_k x\}$ mod 1.

Received by the editor August 4, 2006, and in revised form January 30, 2007.

2000 *Mathematics Subject Classification.* Primary 42C15, 42A55, 42A61, 30B50, 11K38, 60G50.

Key words and phrases. Almost everywhere convergence, mean convergence, lacunary series, Dirichlet series, martingales, quasi-orthogonality, random trigonometric series, random walks, discrepancy.

The first author was supported by OTKA grants K 61052, K 67961 and FWF grant S9603-N13.

Introduction

Throughout this paper $\mathcal{N} = \{n_k, k \geq 1\}$ denotes an increasing sequence of positive integers, and $\mathbf{c} = \{c_k, k \geq 1\}$ some element of ℓ^2. Let $\mathbf{T} = [0,1) = \mathbf{R}/\mathbf{Z}$ be the circle equipped with normalized Lebesgue measure λ, and let $f : \mathbf{T} \to \mathbf{R}$ be a Borel-measurable function. With these quantities in hand, one can formally define the series

$$\sum_{k=1}^{\infty} c_k f(n_k x). \tag{1}$$

The aim of this paper is to study under which conditions the series (1) defines an element of $L^2(\mathbf{T})$ or converges for almost all x in $\mathbf{T}$. We are thus going to investigate the convergence problem of the sequence of partial sums

$$S_N^{\mathcal{N}}(\mathbf{c}, f) = \sum_{k=1}^{N} c_k f(n_k x), \qquad N = 1, 2, \ldots$$

in mean (namely in the space $L^2(\mathbf{T})$) or for almost all x in $\mathbf{T}$. In the trigonometric case this problem has been one of the central problems of harmonic analysis, investigated intensively from the 1920's, culminating in the celebrated theorem of Carleson [8], stating the almost everywhere convergence of the series $\sum_{k=1}^{\infty} c_k \sin 2\pi kx$, $\sum_{k=1}^{\infty} c_k \cos 2\pi kx$ for all $\mathbf{c} \in \ell^2$. Starting from the 1930's, there has been also considerable interest in the convergence properties of the series (1) for general $f \in L^2(\mathbf{T})$, but the existing results are, even today, much less complete than in the trigonometric case. As it turned out, for general f the behavior of the series (1) is radically different from the trigonometric case: the terms of the series are usually far from orthogonal and the convergence properties of the sum depend sensitively on the coefficient sequence (c_k), the analytic properties of f, the growth speed and most importantly on the number-theoretic properties of the sequence n_k. As a result of the 'interference' between the behavior of the Fourier coefficients of f and the arithmetic properties of n_k, even the asymptotic evaluation of the integral

$$\int_{\mathbf{T}} \left(\sum_{k=1}^{N} c_k f(n_k x) \right)^2 dx \tag{2}$$

is generally a hard problem. The first insight into the nature of this integral was given by Wintner [54] who showed that if

$$f \in L^2(\mathbf{T}), \qquad f \sim \sum_{k=1}^{\infty} (a_k \cos 2\pi kx + b_k \sin 2\pi kx),$$

then the system $\{f(kx),\, k \geq 1\}$ is quasi-orthogonal, i.e. there exists a constant $K > 0$ such that any real $\{c_k, 1 \leq k \leq N\}$ we have

$$\int_{\mathbf{T}} \left(\sum_{k=1}^{N} c_k f(kx) \right)^2 dx \leq K \left(\sum_{k=1}^{N} c_k^2 \right)$$

if and only if the Dirichlet series

$$\sum_{n=1}^{\infty} a_n n^{-s}, \quad \text{and} \quad \sum_{n=1}^{\infty} b_n n^{-s} \tag{3}$$

are regular and bounded in the half-plane $\Re(s) > 0$. The last property is therefore necessary and sufficient for the convergence in $L^2(\mathbf{T})$ norm of the series $\sum_{k=1}^{\infty} c_k f(kx)$ for all coefficient sequences $\mathbf{c} \in \ell^2$. Unfortunately, there exist no similarly complete results for the L^2 convergence of the series (1) for general (n_k) and the a.e. convergence problem of the series is unsolved even for $n_k = k$. Specifically, it is unknown for which $f \in L^2(\mathbf{T})$ the series (1) converges a.e. for all $\mathbf{c} \in \ell^2$ ("Carleson" behavior) and even in the case of simple function classes such as $C(\mathbf{T})$, $\mathrm{Lip}_\alpha(\mathbf{T})$, $\mathrm{BV}(\mathbf{T})$, the existing a.e. convergence criteria for $\sum c_k f(kx)$ are far from optimal. The difficulties in this field are well illustrated by the long history of the Khinchin conjecture. Note that if $\sum c_k f(kx)$ converges a.e. for all $\mathbf{c} \in \ell^2$, then choosing $c_k = 1/k$ and using the Kronecker lemma it follows that

$$\lim_{N\to\infty} \frac{1}{N} \sum_{k=1}^{N} f(kx) = 0 \qquad \text{a.e.} \tag{4}$$

Khinchin (1923) formulated the conjecture that $f \in L^1(\mathbf{T})$, $\int_{\mathbf{T}} f(t)dt = 0$ imply (4), a problem which remained open for almost 50 years until Marstrand [34] solved it in the negative. At about the same time, Nikishin [35] constructed the first continuous function $f \in C(\mathbf{T})$ such that $\sum c_k f(kx)$ diverges a.e. for some $\mathbf{c} \in \ell^2$. Despite the profound work of Bourgain [7] connecting the problem with metric entropy behavior, we have no characterization of the class $\mathcal{A}$ of functions f for which (4) holds. The analogous problem for

$$\lim_{N\to\infty} \frac{1}{N} \sum_{k=1}^{N} f(n_k x) = 0 \qquad \text{a.e.}$$

is even more difficult, due to the effect of the arithmetic properties of n_k.

In view of the great difficulties outlined above, one cannot hope for precise norm or a.e. convergence criteria for (1) for general $f \in L^2$ and arbitrary n_k. There exist, however, interesting and important results in certain special situations, in particular for gap series and smooth functions f. Kac [27] proved that if $f \in \mathrm{Lip}(\alpha)$, $\alpha > 0$ and (n_k) satisfies the Hadamard gap condition

$$n_{k+1}/n_k \geq q > 1 \qquad (k = 1, 2, \dots) \tag{5}$$

then the series (1) converges a.e. provided $\mathbf{c} \in \ell^2$. He also showed (see [28]) that in the case $n_k = 2^k$ and under suitable smoothness conditions on f, the sequence $f(n_k x)$ satisfies the central limit theorem, i.e.

$$\lim_{N\to\infty} \lambda\left\{x \in (0,1) : \sum_{k\leq N} f(n_k x) \leq t\sigma\sqrt{N}\right\} = (2\pi)^{-1/2} \int_{-\infty}^{t} e^{-u^2/2} du \tag{6}$$

provided

$$\sigma^2 := \int_0^1 f^2(t)dt + 2\sum_{k=1}^{\infty} \int_0^1 f(t) f(2^k t)dt \neq 0.$$

These results show that under (5) $f(n_k x)$ behaves like a sequence of independent random variables, a fact that explains the nice convergence properties of the series (1) in view of Kolmogorov's three series criterion for independent r.v.'s. As a matter of fact, (6) is not valid for all (n_k) satisfying the Hadamard gap condition (5); it fails, for example for $n_k = 2^k - 1$, as was shown by Erdős and Fortet (see Kac [29], p. 646). Still, there is enough independence in the structure of $f(n_k x)$ to make the a.e. convergence problem of (1) manageable. Whether "true" independence for $f(n_k x)$ holds (e.g. if the CLT (6) is valid) depends again on the number-theoretic properties of (n_k); for a study of the CLT see Gaposhkin [17].

Another situation when the a.e. convergence behavior of (1) is completely understood is the case when f is very smooth. Gaposhkin [16] proved that if the Fourier series of f is absolutely convergent, then (1) converges a.e. for any increasing sequence (n_k) of positive integers and any $\mathbf{c} \in \ell^2$. In particular, this is the case if $f \in \mathrm{Lip}\,(\alpha)$, $\alpha > 1/2$. The Lipschitz condition here is optimal: as it was shown by Berkes [4], for any positive sequence $\varepsilon_k \to 0$ there exists $f \in \mathrm{Lip}\,(1/2)$, $\mathbf{c} \in \ell^2$ and a sequence (n_k) of positive integers satisfying

$$n_{k+1}/n_k \geq 1 + \varepsilon_k \qquad (k = 1, 2, \dots)$$

such that (1) diverges almost everywhere. This counterexample also shows that the Hadamard gap condition (5) in Kac's convergence result above is best possible.

The previously formulated results describe the known cases when the a.e. convergence behavior of (1) is completely understood. Weakening the Hadamard gap condition (5) or the absolute convergence condition on the Fourier series of f leads to a new situation and a completely different convergence behavior of (1), as examples show. The purpose of the present paper is to give a detailed analysis of the probabilistic and harmonic structure of $f(n_k x)$ in the general case which will enable us to give satisfactory a.e. convergence criteria for (1) for a number of important classes of functions f and sequences (n_k). Among others, we will be interested in

(a) The *sub-lacunary* case, i.e. when $n_{k+1}/n_k \to 1$, but its speed of convergence to 1 is restricted by the condition

$$n_{k+1}/n_k \geq 1 + ck^{-\gamma} \qquad (k \geq k_0), \qquad 0 < \gamma < 1 \tag{7}$$

(b) "Nice" function classes like $\mathrm{Lip}_\alpha(\mathbf{T})$, $\mathrm{BV}(\mathbf{T})$, or functions f with sufficiently rapidly converging Fourier series.

(c) Sequences (n_k) with "nice" arithmetic properties.

Under the sub-lacunary condition (7) the central limit theorem (6) is generally false, but we will show that for $\gamma < 1/2$ $f(n_k x)$ has certain martingale properties enabling us to prove that the almost everywhere and L^2 convergence of (1) are equivalent, an important property resembling the classical behavior of sums $\sum X_k$ of independent random variables. Consequently, under (7) for $\gamma < 1/2$ the boundedness and regularity of the Dirichlet series in (3) for $\Re(s) > 0$ is sufficient for the a.e. convergence of (1) for all $\mathbf{c} \in \ell^2$ subject to minor regularity conditions. Conversely, in Chapter 3 we will show that if this boundedness condition is not satisfied, then there exists a $\mathbf{c} \in \ell^2$ and a sequence (n_k) of integers satisfying (7) for all $\gamma < 1/2$ such that (1) diverges a.e. In the absence of the boundedness condition for the Dirichlet series in (3), the almost everywhere (and even mean) convergence of (1) require stronger conditions than $\sum_{k=1}^{\infty} c_k^2 < \infty$ and in Chapter 2 we will give sufficient criteria for this behavior. For example, we will show that (1) converges a.e. provided $f \in \mathrm{Lip}\,(\alpha)$, $\alpha > 0$ and

$$\lim_{R\to\infty} \Big(\sum_{k>R} c_k^2\Big)^{1/2} \Big(\sum_{k>R} n_k^{-2}\Big)^{1/2} \Big(\sum_{k\le R} n_k^{\alpha}\Big)^{1/\alpha} = 0 \tag{8}$$

or

$$\sum_{k=1}^{\infty} c_k^2 \omega(k) < \infty \quad \text{with} \quad \omega(j) := \max\Big(\sum_{1\le \ell\le j} (n_\ell/n_j)^{\alpha}, \sum_{\ell\ge j} (n_j/n_\ell)^{\alpha} \Big). \tag{9}$$

If (n_k) grows almost exponentially, both criteria are close to $\sum_{k=1}^{\infty} c_k^2 < +\infty$, and thus Kac's theorem is "almost" valid in this situation.

The value $\gamma = 1/2$ in (7) is critical: for $\gamma < 1/2$ the CLT (6) still holds for $f(x) = \sin 2\pi x$, $f(x) = \cos 2\pi x$ and this breaks down for $\gamma = 1/2$. (See Erdős [10], cf. also Berkes [3] for more information on this point.) Beyond $\gamma = 1/2$ the behavior of the sums $\sum_{k\le N} e^{2\pi i n_k x}$ becomes pathological and the connection with independent random variables is lost. In this domain arithmetical behavior takes over: the convergence properties of the series $\sum_{k=1}^{\infty} c_k f(n_k x)$ are determined not any more by the growth speed of (n_k), but its number-theoretic properties. To have an idea what type of results one can expect here, note that by a well known observation in the metric theory of Diophantine approximation (see e.g. [33], p. 170), for the function $f(x) = x - [x] - 1/2$ we have

$$\int_{\mathbf{T}} f(n_k x) f(n_l x) dx = \langle n_k, n_l \rangle,$$

where $\langle m, n\rangle = (m,n)/[m,n]$ is the ratio of the greatest common divisor and least common multiple of m and n. This shows that the behavior of the integral (2) and consequently the L^2 convergence of the series (1) are closely related to the norm of the quadratic form $\sum_{k,l=1}^{N} \langle n_k, n_l\rangle x_k x_l$. This leads easily to sufficient conditions for the norm convergence of (1) and using the methods of classical Rademacher–Mensov theory of orthogonal series, also to a.e. convergence criteria. Unfortunately, this method does not extend for general $f \in L^2$ (as is illustrated by Wintner's theorem), and we will use other methods to handle the integral (2). In Chapter 2 we will give several convergence results for (1) with number-theoretic character. A typical result is Theorem 2.4 stating the a.e. convergence of (1) under the condition $\sum_{k=1}^{\infty} c_k^2 (\log k)^2 \lambda_k < \infty$ where

$$\lambda_N = \sup_{1\le h\le N} C_h \sum_{h\le k\le N} \frac{(n_h, n_k)}{n_k} L\Big(\frac{(n_h, n_k) C_k}{n_h}\Big) \tag{10}$$

and C_h is a sequence determined by the convergence speed of the Fourier series of f and $L(x) = \log(x \vee 1)$. Several variants and corollaries of this results are also given, see the corollaries after Theorem 2.5. Particularly simple results are obtained in the case when $n_k = k^r$, $r \ge 2$, when the n_k are coprimes or if the n_k are uniformly distributed in residue classes mod d in the sense of Corollary 2.4.

Our arithmetic convergence criteria in Chapter 2 involving the quantity (10) apply in the case $n_k = k$, but the so obtained results are far from optimal. To deal with this important case, we will use a different approach: we extend Gaposhkin's approximation technique developed for functions $f \in \mathrm{Lip}\,(\alpha)$, $\alpha > 1/2$ to yield information on the connection between the mean convergence properties of the Fourier series of f and a.e. convergence of (1). Specifically, letting

$$S_N(x) = \sum_{k=1}^{N} c_k f(kx)$$

we will give in Theorem 2.6 a maximal inequality for $S_N(x)$ in terms of (c_k) and the Fourier coefficients of f. This estimate is similar to Hunt's inequality [25] and leads to various interesting corollaries regarding the convergence of (1) for special classes of f such as Lip (α), BV, etc.; see Corollaries 2.7–2.9.

In Chapter 3 we will give a number of counterexamples showing the limits of "nice" convergence behavior of (1). We will show, in particular, that the sufficient convergence criteria for the classes Lip, BV obtained in Chapter 2 do not extend for the class L^∞; in fact, with the exception of the trivial case $\sum_{k=1}^\infty |c_k| < \infty$, the convergence behavior of the series (1) for general $f \in L^\infty$ is bad. No condition of the type $\sum_{k=1}^\infty c_k^2 \omega(k) < \infty$ implies the a.e. convergence of (1) for all bounded measurable f even if n_k grows exponentially fast or if n_k have a nice arithmetic structure, e.g. if all the n_k are primes.

As we have seen, the convergence behavior of the series (1) depends strongly on the sequence (n_k), and thus it is natural to ask if there is a "typical" behavior, valid for "most" sequences (n_k). Of course, to give a meaning to the phrase "most sequences", one should define a natural probability measure μ over the class of increasing integer sequences (n_k) and define "typical behavior" with respect to μ. Equivalently, we should choose $n_k = n_k(\omega)$ at random and ask what is the typical convergence behavior of the series (1) with probability one, i.e. for almost every ω. Certainly, the most natural choice of a random structure for (n_k) is an increasing random walk, i.e. $n_k = X_1 + \ldots + X_k$, where X_k are i.i.d. positive integer valued random variables defined on some probability space $(\Omega, \mathcal{A}, \mathbf{P})$. Actually, we will not assume that X_k are integer valued, thus permitting non-integral sequences (n_k) as well, stipulating only that $n_k \to \infty$ almost surely. In Chapter 4 we will investigate this question and show that there is indeed a typical convergence behavior of the series (1), but it is different for integral valued or continuously distributed X_k. Theorem 4.2 shows that if X_k have a bounded density, the series (1) satisfies, with probability one, the analogue of Carleson's theorem: for almost every ω, $\sum_{k=1}^\infty c_k f(n_k(\omega)x)$ converges for almost every x provided $\mathbf{c} \in \ell^2$. If the X_k are integer valued, the situation is more complicated. The proof of Theorem 4.3 shows that the convergence of (1) is equivalent to that of $\sum_{j=1}^\infty d_j f(jx)$ where d_j are random coefficients explicitly computable from the coefficients c_k and the random variables X_k. In particular, if $\sum c_k^2 < \infty$, then $\sum d_j^2 < \infty$ a.s. This gives very precise information on the L^2 convergence of (1): if the Dirichlet series in (3) are bounded and regular for $\Re(s) > 0$, then for almost every ω, $\sum_{k=1}^\infty c_k f(n_k(\omega)x)$ converges in L^2 norm for all $\mathbf{c} \in \ell^2$. Convergence almost everywhere (with respect to x) is a more difficult matter, where, for example, the sufficient criteria for the a.e. convergence of $\sum_{k=1}^\infty c_k f(kx)$ obtained in Chapter 2 can be utilized.

A remarkable phenomenon holds in the trigonometric case $f(x) = e^{2\pi ix}$. By Carleson's theorem, for any (deterministic) increasing sequence (n_k) of positive integers, the series $\sum_{k=1}^{\infty} c_k e^{2\pi i n_k x}$ converges a.e. provided $\mathbf{c} \in \ell^2$. While this is a very deep result, a simple elementary argument in Chapter 4 will show that if (n_k) is any nondegenerate random walk (i.e. if the generating random variables X_k are not constant), then with probability one, $\sum_{k=1}^{\infty} c_k e^{2\pi i n_k x}$ converges for almost every x, provided $\mathbf{c} \in \ell^2$. Actually, we need not assume that the X_k are integer valued, or even that they are positive. The result remains valid even if the random walk (n_k) is recurrent, e.g. if $n_k = 0$ for infinitely many k or (n_k) is dense on the real line. See Chapter 4 for the explanation of this curious phenomenon.

The previous results show that for random (n_k), the sequence $f(n_k x)$ has much nicer properties than in the deterministic case. In Chapter 5, we illustrate this phenomenon from another side, studying the discrepancy of the sequence $\{n_k x\}$, where $\{\cdot\}$ denotes fractional part. Given an infinite sequence $\mathbf{s} = (s_1, s_2, \dots)$ in $[0, 1)$, the discrepancy $D_N(\mathbf{s})$ of $\mathbf{s}$ is defined by

$$D_N(\mathbf{s}) = \sup_{0 \le a < b \le 1} \frac{1}{N} \left| \sum_{\substack{n=1 \\ s_n \in [a,b)}}^{N} 1 - N(b-a) \right|.$$

We call $\mathbf{s}$ uniformly distributed in the Weyl sense if $D_N(\mathbf{s}) \to 0$ as $N \to \infty$. By a classical result of Weyl [52], for any increasing sequence (n_k) of positive integers, $\{n_k x\}$ is uniformly distributed for almost every $x \in \mathbf{R}$. Unfortunately, with the exception of lacunary sequences and a few special subexponential sequences (n_k) (see Kesten [30], Philipp [40], [41]) no precise estimates for the discrepancy of $\{n_k x\}$ are known. Complementing the results in Chapter 4, in Chapter 5 we give discrepancy estimates for random (n_k). Again, there is a substantial difference between the cases when the i.i.d. random variables X_k generating (n_k) are discrete or continuous. In the case when X_k are absolutely continuous, Schatte [45] showed that with probability one, the precise order of magnitude of the discrepancy of $\{n_k x\}$ is $O((\log\log N/N)^{1/2})$ for almost every x. We will study the case of integer valued n_k, which is more complicated, and give various bounds for the discrepancy, exhibiting phenomena very different from the deterministic case.

CHAPTER 1

Mean convergence

In this chapter we will investigate the mean convergence of the series

$$\sum_{k=1}^{\infty} c_k f(n_k x). \tag{1.1}$$

As we already noted, this is equivalent to the asymptotic evaluation of the integral

$$\int_{\mathbf{T}} \left(\sum_{k=1}^{N} c_k f(n_k x) \right)^2 dx. \tag{1.2}$$

Such a study is clearly a prerequisite to the almost everywhere convergence of the series (1.1) and some facts and estimates collected in this chapter will be used repeatedly in later chapters. We start with formulating Wintner's fundamental theorem (see [54]) concerning the case $n_k = k$.

THEOREM A. *Let $f \in L^2(\mathbf{T})$ with $\int_{\mathbf{T}} f(t)dt = 0$ and with Fourier series*

$$f \sim \sum_{k=1}^{\infty} (a_k \cos 2\pi kx + b_k \sin 2\pi kx). \tag{1.3}$$

Then the following statements are equivalent:

(a) *The series $\sum_{k=1}^{\infty} c_k f(kx)$ converges in $L^2(\mathbf{T})$ for any $\mathbf{c} \in \ell^2$.*

(b) *There exists a constant $K > 0$ such that for any $n \geq 1$ and any real $\{c_k, 1 \leq k \leq n\}$ we have*

$$\int_{\mathbf{T}} \left(\sum_{k=1}^{n} c_k f(kx) \right)^2 dx \leq K \left(\sum_{k=1}^{n} c_k^2 \right).$$

(c) *The infinite matrix*

$$\int_{\mathbf{T}} f(kt) f(\ell t) dt \qquad\qquad (k, \ell = 1, 2 \ldots)$$

defines a bounded operator on ℓ^2.

(d) *The Dirichlet series*

$$\sum_{n=1}^{\infty} a_n n^{-s}, \quad \text{and} \quad \sum_{n=1}^{\infty} b_n n^{-s} \tag{1.4}$$

are regular and bounded in the half-plane $\Re(s) > 0$.

The basic ingredient of Wintner's proof is Toeplitz's criterion [47] for the ℓ^2 boundedness of so called "D-matrices" in terms of Dirichlet series. The connection with the convergence problem in Theorem A is established by the Möbius transformation; see Wintner [54] for the details. As a comparison, note that the assumption $f \in L^2(\mathbf{T})$, i.e. $\sum_{k=1}^{\infty}(a_k^2+b_k^2)<\infty$ implies only that the sums in (1.4) are absolutely convergent for $\Re(s)>1/2$.

Clearly, condition (a) of Theorem A implies that $\sum_{k=1}^{\infty} c_k f(kx)$ converges in measure for any $\mathbf{c} \in \ell^2$. By a remarkable result of Nikishin [37, Theorem 12], the converse is also true, i.e. the convergence theory of $\sum_{k=1}^{\infty} c_k f(kx)$ is the same for L^2 convergence and convergence in measure.

Note that if condition (d) of Theorem A is not satisfied, the series $\sum_{k=1}^{\infty} c_k f(kx)$ can still converge in $L^2(\mathbf{T})$ for a large class of coefficient sequences (c_k). For example, Wintner [54] noted that if $f \in L^2(\mathbf{T})$, $\int_{\mathbf{T}} f(t)dt=0$ with Fourier series (1.3) then $\sum_{k=1}^{\infty} c_k f(kx)$ converges in $L^2(\mathbf{T})$ if $a_k=O(k^{-\gamma})$, $b_k=O(k^{-\gamma})$, $c_k=O(k^{-\gamma})$ for some $\gamma>1/2$. Note that here the assumptions made on the Fourier coefficients of f do not in general imply the boundedness of the Dirichlet series in (1.4) and accordingly, the assumption made on the coefficient sequence (c_k) is stronger than $\mathbf{c} \in \ell^2$.

An application of the last remark is the series

$$\sum_{k=1}^{\infty} \frac{\psi(kx+\frac{1}{2})}{k}, \tag{1.5}$$

where

$$\psi(x)=\begin{cases} x-[x]-\frac{1}{2}, & \text{if } x \neq [x], \\ 0, & \text{if } x=[x]. \end{cases}$$

This example has considerable historical interest, since it was used by Riemann [42, p. 263] to illustrate the limitations of his own integration theory. He showed that both (1.5) and the trigonometric sum

$$\sum_{n=1}^{\infty} \frac{c(n)}{n} \sin 2\pi n x, \tag{1.6}$$

where

$$c(n)=\sum_{d|n}(-1)^d \tag{1.7}$$

converge if x is rational, and to the same limit. Moreover, he observed that the function defined by these series on the set of rational numbers is unbounded on any interval, and thus (1.6) cannot be the Fourier series of its sum in the Riemann sense. From the remarks after Theorem A it follows that (1.5) converges in $L^2(\mathbf{T})$ and Wintner showed in [53] that its sum belongs to $L^p(\mathbf{T})$ for any $p>1$ and has (1.6)–(1.7) as its Fourier series in the Lebesgue sense.

Let us recall that a sequence of vectors $(x_n)_{n\geq 1}$ in a Hilbert space H is called a Riesz basis if every $x \in H$ can be expanded as $f=\sum_{n=1}^{\infty} a_n x_n$ and there exist positive constants C_1, C_2 such that for any $n \geq 1$ and any real sequence $(a_k)_{1\leq k\leq n}$ we have

$$C_1\left(\sum_{k=1}^{n} a_k^2\right) \leq \left\|\sum_{k=1}^{n} a_k x_k\right\|^2 \leq C_2\left(\sum_{k=1}^{n} a_k^2\right).$$

Hedenmalm, Lindquist and Seip proved (see [23], [24]) that if

$$f \in L^2(\mathbf{T}), \quad f(t) \sim \sum_{k=1}^{\infty} d_k \sin 2\pi k t,$$

then $\{f(nx), n \geq 1\}$ is a Riesz basis in $L^2(\mathbf{T})$ if and only if the Dirichlet series $\sum_{n=1}^{\infty} d_n n^{-s}$ is analytic and bounded away from 0 and ∞ in the whole right half-plane $\Re(s) > 0$, i.e.

$$\delta \leq \left| \sum_{n=1}^{\infty} d_n n^{-\sigma-it} \right| \leq \Delta \quad \text{for all } \sigma > 0,\ t \in \mathbf{R}$$

with some positive constants δ and Δ. See also the preceding work of Gosselin and Neuwirth [20], as well as the article by Ginsberg, Neuwirth and Newman [19].

So far, we have considered the convergence in mean problem in the case $\mathcal{N} = \mathbf{N}$. In the general case the existing results in the literature are much less complete, due to number-theoretic difficulties. Recall the notation

$$\langle a, b \rangle = \frac{(a,b)}{[a,b]},$$

where (a,b) and $[a,b]$ denote the greatest common divisor resp. least common multiple of the positive integers a and b. The following theorem is an easy consequence of results of Wintner [54].

THEOREM B. *Let $f \in L^2(\mathbf{T})$ with $\int_{\mathbf{T}} f(t)dt = 0$ and Fourier series*

$$f \sim \sum_{k=1}^{\infty} \big(a_k \cos 2\pi k t + b_k \sin 2\pi k t\big),$$

where $a_k = O(k^{-\alpha})$, $b_k = O(k^{-\alpha})$, $\alpha > 1/2$. Let (n_k) be an increasing sequence of positive integers and (c_k) a real coefficient sequence. Then $\sum_{k=1}^{\infty} c_k f(n_k x)$ converges in the mean provided

$$\sum_{k,l=1}^{\infty} |c_k||c_l| \langle n_k, n_l \rangle^{\alpha} < \infty. \tag{1.8}$$

Note that the assumptions made on the Fourier coefficients of f in Theorem B do not imply condition (d) of Theorem A. For example, if $f \sim \sum_{k=1}^{\infty} k^{-1} \sin 2\pi k x$, then $\sum_{k=1}^{\infty} a_n n^{-s} = \zeta(1+s)$ for $\Re(s) > 0$, which tends to ∞ if $s \to +0$.

To prove Theorem B, it suffices to consider the case when the Fourier series of f is a pure sine or cosine series. Now if $f \sim \sum_{k=1}^{\infty} a_k \cos 2\pi k t$, then by the assumption on the a_k and relation (52) of Wintner [54] we have for any positive integers i, j

$$\int_{\mathbf{T}} f(it) f(jt) dt = \frac{1}{2} \sum_{h=1}^{\infty} a_{hi/(i,j)} a_{hj/(i,j)} \leq C_1 \langle i, j \rangle^{\alpha} \sum_{h=1}^{\infty} h^{-2\alpha} \leq C_2 \langle i, j \rangle^{\alpha}$$

for some constants C_1, C_2 and thus

$$\int_{\mathbf{T}} \left(\sum_{k=m}^{n} c_k f(n_k t) \right)^2 dt \leq C_2 \sum_{m \leq k, l \leq n} |c_k||c_l| \langle n_k, n_l \rangle^{\alpha} \tag{1.9}$$

Hence the mean convergence of $\sum_{k=1}^{\infty} c_k f(n_k x)$ follows from (1.8).

In particular, under the assumptions made on f in Theorem B, $\sum_{k=1}^{\infty} c_k f(n_k x)$ converges in $L^2(\mathbf{T})$ norm for any $\mathbf{c} \in \ell^2$ provided the quadratic form

$$\sum_{k,l=1}^{\infty} \langle n_k, n_l \rangle^{\alpha} x_k x_l \tag{1.10}$$

is bounded, i.e. there exists a constant $A > 0$ such that

$$\left| \sum_{k,l=1}^{N} \langle n_k, n_l \rangle^{\alpha} x_k x_l \right| \leq A \sum_{n=1}^{N} x_n^2 \tag{1.11}$$

for any $N \geq 1$ and any real $x_1, \ldots, x_N$. This is equivalent, in turn, to the fact that the matrix $\langle n_k, n_l \rangle^{\alpha}$ $(k, l = 1, 2, \ldots)$ defines a bounded operator on ℓ^2. In the case $n_k = k$ this holds if and only if $\alpha > 1$, as it is shown in [54], pp. 577–578. Also, if the n_k are coprimes, then $\langle n_k, n_l \rangle = (n_k n_l)^{-1}$ and thus by Cauchy's inequality, (1.11) is satisfied if $\sum_{k=1}^{\infty} n_k^{-2\alpha} < \infty$. For general (n_k), a sufficient condition for (1.11) is (see the proof of Lemma 7.4.3 in Weber [48], p. 139),

$$\sup_{k \geq 1} \sum_{l \geq 1} \langle n_k, n_l \rangle^{\alpha} < \infty. \tag{1.12}$$

Unfortunately, computing the order of magnitude of the sums in (1.11), (1.12) for general (n_k) is a difficult number-theoretic problem. In a profound paper, Gál [12] showed that for any increasing (n_k) we have

$$\sum_{k,l=1}^{N} \langle n_k, n_l \rangle \leq CN (\log \log N)^2$$

and he constructed an (n_k) for which the bound $CN(\log \log N)^2$ is actually reached. For this sequence (n_k), relation (1.11) clearly fails for $x_1 = \cdots = x_N = 1$. No sharp estimate for the left-hand side of (1.11) is known for general $x_1, \ldots, x_N$.

The following theorem gives mean convergence criteria for $\sum_{k=1}^{\infty} c_k f(n_k x)$ in the case when (1.12) is not satisfied.

THEOREM 1.1. *Let $f \in L^2(\mathbf{T})$ with $\int_{\mathbf{T}} f(t) dt = 0$ and Fourier series*

$$f(t) \sim \sum_{k=1}^{\infty} (a_k \cos 2\pi k t + b_k \sin 2\pi k t)$$

where $a_k = O(k^{-\alpha})$, $b_k = O(k^{-\alpha})$, $\alpha > 1/2$. Let (n_k) be an increasing sequence of positive integers and let (λ_n) be a positive nondecreasing sequence such that $\lambda_{2n}/\lambda_n = O(1)$ and

$$\sup_{1 \leq k \leq N} \sum_{l=1}^{N} \langle n_k, n_l \rangle^{\alpha} \leq \lambda_N. \tag{1.13}$$

Then $\sum_{k=1}^{\infty} c_k f(n_k x)$ converges in $L^2(\mathbf{T})$ norm provided

$$\sum_{k=1}^{\infty} c_k^2 (\log k)^{\gamma} \lambda_k < \infty \quad \textit{for some } \gamma > 1. \tag{1.14}$$

Note that in the case when $\lambda_N = O(1)$, $\sum_{k=1}^{\infty} c_k f(n_k x)$ converges in the mean provided $\sum_{k=1} c_k^2 < \infty$ (see above), but condition (1.14) specialized to this case gives a more stringent condition. This is due to the fairly crude estimates we use for the quadratic form appearing in the argument.

We formulate a few corollaries of Theorem 1.1, Theorem B and the arguments following the proof of Theorem B.

COROLLARY 1.1. *Assume f satisfies the assumptions of Theorem 1.1 and let $n_k = k^r$ where $r \geq 2$ is an integer. Then $\sum_{k=1}^{\infty} c_k f(n_k x)$ converges in the mean provided $\sum_{k=1}^{\infty} c_k^2 < \infty$.*

As we noted above, the assumptions made on the Fourier coefficients of f in Corollary 1.1 do not imply condition (d) of Theorem A, but $\sum_{k=1}^{\infty} c_k f(n_k x)$ still converges for all $\mathbf{c} \in \ell^2$. This is due to the speed and nice number-theoretic properties of n_k.

COROLLARY 1.2. *Assume f satisfies the assumptions of Theorem 1.1. Then the series $\sum_{k=1}^{\infty} c_k f(kx)$ converges in the mean provided*

$$\sum_{k=1}^{\infty} c_k^2 (\log k)^{3+\varepsilon} < \infty, \qquad \text{if } \alpha = 1$$

and

$$\sum_{k=1}^{\infty} c_k^2 (\log k)^{1+\varepsilon} k^{1-\alpha} < \infty, \qquad \text{if } \alpha < 1.$$

Note that the case $\alpha > 1$ is uninteresting: in this case $\sum_{k=1}^{\infty} (|a_k| + |b_k|) < \infty$ and thus

$$\begin{aligned}\left\| \sum_{k=1}^{N} c_k f(n_k x) \right\| &\leq \sum_{j=1}^{\infty} |a_j| \left\| \sum_{k=1}^{N} c_k \cos 2\pi j n_k x \right\| + \sum_{j=1}^{\infty} |b_j| \left\| \sum_{k=1}^{N} c_k \sin 2\pi j n_k x \right\| \\ &\leq C \left(\sum_{k=1}^{N} c_k^2 \right)^{1/2}\end{aligned}$$

for some constant C and thus $\sum_{k=1}^{\infty} c_k f(n_k x)$ converges in the mean for any $\mathbf{c} \in \ell^2$. (Actually, the series converges almost everywhere also, see Gaposhkin [16].)

COROLLARY 1.3. *Let f satisfy the assumptions of Theorem 1.1 with $\alpha = 1$ and let (n_k) be a sequence of integers such that for any $d \geq 1$ we have $\sum_{d|n_k} n_k^{-1} \leq A/d$ with an absolute constant A. Then $\sum_{k=1}^{\infty} c_k f(n_k x)$ converges in the mean provided*

$$\sum_{k=1}^{\infty} c_k^2 (\log k)^{\gamma} (\log n_k) < \infty, \qquad \gamma > 1. \tag{1.15}$$

The condition $\sum_{d|n_k} n_k^{-1} \leq A/d$ is satisfied if the sequence (n_k) is roughly uniformly distributed among the residue classes mod d. It is also satisfied if the n_k are coprimes, since then for any d the sum $\sum_{d|n_k} n_k^{-1}$ contains at most one term. Further, this condition is satisfied for $n_k = k^r$, $r \geq 2$ but in this case Corollary 1.1 gives a better result. We see again that the number-theoretic properties of n_k play a crucial role in the convergence behavior of $\sum_{k=1}^\infty c_k f(n_k x)$, which can be anticipated from Theorem B. If (n_k) grows with a polynomial speed, then $\log n_k = O(\log k)$ and thus the convergence condition (1.15) reduces to $\sum c_k^2 (\log k)^{2+\varepsilon} < \infty$.

PROOF OF THEOREM 1.1. By assumption (1.13), relation (1.9) and the Cauchy–Schwarz inequality we have

$$\begin{aligned}\int_{\mathbf{T}} \left(\sum_{k=1}^N c_k f(n_k t)\right)^2 dt &\le C_2 \sum_{k,l=1}^N |c_k||c_l| \langle n_k, n_l\rangle^\alpha \\ &\le C_2 \sum_{k,l=1}^N \frac{1}{2}(c_k^2 + c_l^2)\langle n_k, n_l\rangle^\alpha \le C_2 \lambda_N \left(\sum_{k=1}^N c_k^2\right)\end{aligned} \tag{1.16}$$

for any real $c_1, \dots, c_N$. Assume now (1.14) and let $Z_\nu = \sum_{k=2^\nu+1}^{2^{\nu+1}} c_k f(n_k t)$. By the Cauchy-Schwarz inequality we have

$$\left(\sum_{k=2^m+1}^{2^n} c_k f(n_k t)\right)^2 = \left(\sum_{\nu=m}^{n-1} Z_\nu\right)^2 \le \left(\sum_{\nu=m}^{n-1} \nu^\gamma Z_\nu^2\right)\left(\sum_{\nu=1}^{\infty} \nu^{-\gamma}\right). \tag{1.17}$$

Thus with some constant C we have, using (1.16) and $\lambda_{2n}/\lambda_n = O(1)$,

$$\begin{aligned}\int_{\mathbf{T}} \left(\sum_{k=2^m+1}^{2^n} c_k f(n_k t)\right)^2 dt &\le C\left(\sum_{\nu=m}^{n-1} \nu^\gamma \int_{\mathbf{T}} Z_\nu^2 dt\right) \\ &\le C' \sum_{\nu=m}^{n-1} \nu^\gamma \left(\sum_{k=2^\nu+1}^{2^{\nu+1}} c_k^2\right) \lambda_{2^{\nu+1}} \le C'' \sum_{k=2^m+1}^{2^n} c_k^2 (\log k)^\gamma \lambda_k.\end{aligned} \tag{1.18}$$

Here the last expression tends to 0 as $m, n \to \infty$ and since in (1.18) an arbitrary subset of the c_k's can be replaced by 0's, it follows that the L^2 norm of $\sum_{k=i}^j c_k f(n_k x)$ tends to 0 if $i, j \to \infty$, completing the proof of Theorem 1.1. □

Note that in the case $n_k = k^r$, $r \in \mathbf{N}$ we have

$$\langle n_k, n_l\rangle^\alpha = \langle k, l\rangle^{r\alpha}$$

Thus in view of Theorem 1.1, Theorem B and the arguments following Theorem B, for the proof of Corollaries 1.1 and 1.2 it suffices to prove the following

LEMMA 1.1. *Let $\beta > 0$ and*

$$\lambda_N^* = \sup_{1 \le h \le N} \sum_{k=1}^N \langle h, k\rangle^\beta. \tag{1.19}$$

Then $\lambda_N^* = O(1)$, $\lambda_N^* = O(\log^2 N)$ and $\lambda_N^* = O(N^{1-\beta})$ according as $\beta > 1$, $\beta = 1$ or $\beta < 1$.

PROOF. Fix $1 \le h \le N$ and $d|h$ and sum in (1.19) first for those k for which $(h,k)=d$. Then we get

$$\begin{aligned} &\sum_{k\le N,(h,k)=d} \left(\frac{d}{[h,k]}\right)^\beta \\ &= \sum_{k\le N,(h,k)=d} \left(\frac{d^2}{hk}\right)^\beta \le \left(\frac{d}{h}\right)^\beta \sum_{k\le N,d|k} \left(\frac{d}{k}\right)^\beta \le \left(\frac{d}{h}\right)^\beta \sum_{l=1}^{[N/d]} \frac{1}{l^\beta}. \end{aligned} \tag{1.20}$$

For $\beta=1$ the last sum in (1.20) is at most $C^* \log N$ and thus summing for all $d|h$ and noting that the sum of all divisors of h is $\le Ch\log h$, we get the statement of the lemma. For $\beta>1$ the last sum in (1.20) is $O(1)$ and

$$\sum_{d|h} d^\beta \le \sum_{d|h, d\le\sqrt{h}} d^\beta + \sum_{d|h, d\le\sqrt{h}} (h/d)^\beta.$$

Let $\varepsilon>0$. Since the number of divisors of h is $O(h^\varepsilon)$, the first sum on the right-hand side has $O(h^\varepsilon)$ terms and thus this sum is $O(h^{\beta/2+\varepsilon})$; the second sum on the right side is at most $h^\beta \sum_{j=1}^\infty j^{-\beta} = O(h^\beta)$. Choosing ε sufficiently small, we get the statement of the lemma in the case $\beta>1$. Let finally $0<\beta<1$. Then the last expression in (1.20) is

$$\ll \left(\frac{d}{h}\right)^\beta \left(\frac{N}{d}\right)^{1-\beta} = \frac{1}{h^\beta} N^{1-\beta} d^{2\beta-1}$$

and thus the λ_N^* is

$$\ll \sup_{1\le h\le N} \frac{1}{h^\beta} N^{1-\beta} \sum_{d|h} d^{2\beta-1}. \tag{1.21}$$

Let $0<\varepsilon<\min(\beta, 1-\beta)$. Since the number of divisors of h is $O(h^\varepsilon)$, for $\beta \ge 1/2$ the sum in (1.21) is $O(h^{2\beta-1+\varepsilon})$ and thus the expression after the sup in (1.21) is

$$\ll \frac{1}{h^\beta} N^{1-\beta} h^{2\beta-1+\varepsilon} = h^{\beta-1+\varepsilon} N^{1-\beta} \le N^{1-\beta}.$$

If $\beta<1/2$, then the sum in (1.21) is $O(h^\varepsilon)$ and thus the expression after the sup in (1.21) is

$$\ll \frac{1}{h^\beta} N^{1-\beta} h^\varepsilon \le N^{1-\beta}.$$

Thus in both cases the expression in (1.21) is $O(N^{1-\beta})$, and thus the lemma is proved. □

To prove Corollary 1.3, it suffices to show that

$$\sup_{1\le h\le N} \sum_{k=1}^N \frac{(n_h, n_k)}{[n_h, n_k]} \le C \log n_N. \tag{1.22}$$

Fix $1 \le h \le N$ and $d|n_h$ and compute the sum in (1.22) for those $1 \le k \le N$ such that $(n_h, n_k) = d$. This restricted sum clearly cannot exceed, in view of the assumption of Corollary 1.3,

$$\sum_{1\le k\le N, d|n_k} \frac{d^2}{n_h n_k} \le \frac{d^2}{n_h} \frac{A}{d}.$$

Summing for all $d|n_h$, and using the fact that the sum of divisors of n_h is $O(n_h \log n_h)$, we get (1.22).

Our next theorem gives a necessary and sufficient condition for the mean convergence of the series $\sum_{k=1}^{\infty} c_k f(n_k x)$ in terms of the coefficients c_k and the Fourier coefficients of f. Although it is very precise, it is mainly of theoretical interest only since its number-theoretical character makes it difficult to apply in concrete cases.

THEOREM 1.2. *Let $f \in L^2(\mathbf{T})$, $\int_{\mathbf{T}} f(t)dt = 0$ have complex Fourier series $f \sim \sum_{k\in Z, k\neq 0} \varphi_k e_k$ where $e_k(x) = \exp(2\pi i k x)$. Let (n_k) be an increasing sequence of positive integers. Then $\sum_{k=1}^{\infty} c_k f(n_k x)$ converges in $L^2(\mathbf{T})$ norm if and only if the following conditions are fulfilled:*

$$\begin{aligned} &\text{a)} \qquad \lim_{R\to\infty} \sup_{P\geq R} \sum_{|n|>n_R} \Big(\sum_{\substack{n_k|n \\ k\leq P}} \varphi_{n/n_k} c_k\Big)^2 = 0, \\ &\text{b)} \qquad \sum_n \Big(\sum_{n_k|n} \varphi_{n/n_k} c_k\Big)^2 < \infty. \end{aligned} \tag{1.23}$$

If both sequences $(\varphi_n)_{n\in\mathbf{Z}}$ and (c_n) have constant signs, then (1.23a) follows from (1.23b), so that $\sum_{k=1}^{\infty} c_k f(n_k x)$ converges in $L^2(\mathbf{T})$ norm if and only if condition (1.23b) holds. Also, if $\sum_n (\sum_{n_k|n} |\varphi_{n/n_k} c_k|)^2 < \infty$, then $\sum_{k=1}^{\infty} c_k f(n_k x)$ converges in $L^2(\mathbf{T})$ norm.

PROOF. Observe that

$$S_N^{\mathcal{N}}(\mathbf{c}, f) = \sum_n e_n \Big(\sum_{\substack{n_k|n \\ k\leq N}} \varphi_{n/n_k} c_k\Big) = \sum_{|n|\leq n_N} e_n \sum_{n_k|n} \varphi_{n/n_k} c_k + \sum_{|n|>n_N} e_n \sum_{\substack{n_k|n \\ k\leq N}} \varphi_{n/n_k} c_k.$$

Let $M \geq N \geq R$. Then,

$$\begin{aligned} \langle S_N^{\mathcal{N}}(\mathbf{c}, f), S_M^{\mathcal{N}}(\mathbf{c}, f)\rangle &= \sum_n \Big(\sum_{\substack{n_k|n \\ k\leq N}} \varphi_{n/n_k} c_k\Big)\Big(\sum_{\substack{n_k|n \\ k\leq M}} \varphi_{n/n_k} c_k\Big) \\ &= \sum_{|n|\leq n_R} \Big(\sum_{\substack{n_k|n \\ k\leq N}} \varphi_{n/n_k} c_k\Big)^2 + \sum_{|n|>n_R} \Big(\sum_{\substack{n_k|n \\ k\leq N}} \varphi_{n/n_k} c_k\Big)\Big(\sum_{\substack{n_k|n \\ k\leq M}} \varphi_{n/n_k} c_k\Big). \end{aligned}$$

Thus

$$\begin{aligned} &\Big|\langle S_N^{\mathcal{N}}(\mathbf{c}, f), S_M^{\mathcal{N}}(\mathbf{c}, f)\rangle - \sum_{|n|\leq n_R} \Big(\sum_{\substack{n_k|n \\ k\leq N}} \varphi_{n/n_k} c_k\Big)^2\Big| \\ &\qquad \leq \Big[\sum_{|n|>n_R} \Big|\sum_{\substack{n_k|n \\ k\leq N}} \varphi_{n/n_k} c_k\Big|^2\Big]^{\frac{1}{2}} \Big[\sum_{|n|>n_R} \Big|\sum_{\substack{n_k|n \\ k\leq M}} \varphi_{n/n_k} c_k\Big|^2\Big]^{\frac{1}{2}} \\ &\qquad \leq \sup_{P\geq R} \sum_{|n|>n_R} \Big|\sum_{\substack{n_k|n \\ k\leq P}} \varphi_{n/n_k} c_k\Big|^2 \\ &\qquad \to 0, \end{aligned}$$

as R tends to infinity by assumption. Consequently,

$$\lim_{R\to\infty}\sup_{M,N\geq R}\left|\langle S_N^{\mathcal{N}}(\mathbf{c},f),S_M^{\mathcal{N}}(\mathbf{c},f)\rangle-\sum_{|n|\leq n_R}\Big(\sum_{\substack{n_k|n\\k\leq N}}\varphi_{n/n_k}c_k\Big)^2\right|=0.$$

In other words,

$$\lim_{M,N\to\infty}\big\langle S_N^{\mathcal{N}}(\mathbf{c},f),S_M^{\mathcal{N}}(\mathbf{c},f)\big\rangle=A:=\sum_n\Big(\sum_{n_k|n}\varphi_{n/n_k}c_k\Big)^2<\infty.$$

And also

$$\lim_{N\to\infty}\|S_N^{\mathcal{N}}(\mathbf{c},f)\|_2^2=A.$$

These two facts then imply that

$$\lim_{N,M\to\infty}\|S_N^{\mathcal{N}}(\mathbf{c},f)-S_M^{\mathcal{N}}(\mathbf{c},f)\|_2=0,$$

as required.

Conversely, if the sequence $\{S_N^{\mathcal{N}}(\mathbf{c},f),N\geq 1\}$ converges in mean, it is then bounded in mean:

$$\sup_{N\geq 1}\|S_N^{\mathcal{N}}(\mathbf{c},f)\|_2=B<\infty.$$

But as

$$\|S_N^{\mathcal{N}}(\mathbf{c},f)\|_2^2=\sum_{|n|\leq n_N}\Big(\sum_{n_k|n}\varphi_{n/n_k}c_k\Big)^2+\sum_{|n|>n_N}\Big(\sum_{\substack{n_k|n\\k\leq N}}\varphi_{n/n_k}c_k\Big)^2,$$

this implies that $A\leq B$. Now let f^* denote the limit in mean of the sequence $\{S_N^{\mathcal{N}}(\mathbf{c},f),N\geq 1\}$. From

$$\big|\langle f^*,e_n\rangle-\big\langle S_N^{\mathcal{N}}(\mathbf{c},f),e_n\big\rangle\big|\leq\|f^*-S_N^{\mathcal{N}}(\mathbf{c},f)\|_2$$

we deduce

$$\langle f^*,e_n\rangle=\lim_{N\to\infty}\langle S_N^{\mathcal{N}}(\mathbf{c},f),e_n\rangle=\lim_{N\to\infty}\sum_{\substack{n_k|n\\k\leq N}}\varphi_{n/n_k}c_k=\sum_{n_k|n}\varphi_{n/n_k}c_k.$$

Thus $f^*=\sum_{n\in\mathbf{Z}}e_n\sum_{n_k|n}\varphi_{n/n_k}c_k$. Let R be some positive integer and define $H_R=\langle e_n,|n|\leq n_R\rangle$. Let p_R be the projection onto the orthogonal complement ${H_R}^{\perp}$ of H_R. Then,

$$\big|\|p_R(f^*)\|_2-\|p_R(S_N^{\mathcal{N}}(\mathbf{c},f))\|_2\big|\leq\|p_R(f^*)-p_R(S_N^{\mathcal{N}}(\mathbf{c},f))\|_2\leq\|f^*-S_N^{\mathcal{N}}(\mathbf{c},f)\|_2\to 0,$$

as N tend to infinity. Thus,

$$\sup_{N\geq R}\Big|\big\|p_R(f^*)\big\|_2-\big\|p_R(S_N^{\mathcal{N}}(\mathbf{c},f))\big\|_2\Big|\leq\sup_{N\geq R}\big\|f^*-S_N^{\mathcal{N}}(\mathbf{c},f)\big\|_2\to 0,$$

as R tends to infinity. Now, by the triangle inequality,

$$\begin{aligned}\sup_{N\geq R}\left\|p_R(S_N^{\mathcal{N}}(\mathbf{c},f))\right\|_2 &= \sup_{N\geq R}\left[\sum_{|n|>n_R}\Big(\sum_{\substack{n_k|n\\k\leq N}}\varphi_{n/n_k}c_k\Big)^2\right]^{\frac{1}{2}}\\ &\leq \sup_{N\geq R}\left|\left\|p_R(f^*)\right\|_2-\left\|p_R(S_N^{\mathcal{N}}(\mathbf{c},f))\right\|_2\right|+\|p_R(f^*)\|_2\\ &\to 0,\end{aligned}$$

as R tends to infinity. This completes the proof. □

CHAPTER 2

Almost everywhere convergence: sufficient conditions

Let $f \in L^2(\mathbf{T})$ with $\int_{\mathbf{T}} f(t)dt = 0$ and let $\mathcal{N} = \{n_k, k \geq 1\}$ be an increasing sequence of positive integers. Using standard terminology, we call the pair $(f, \mathcal{N})$ (or, equivalently, the sequence $f(n_k x)$) a *convergence system* if for any $\mathbf{c} \in \ell^2$, $\sum_{k=1}^{\infty} c_k f(n_k x)$ converges for almost all $x \in (0,1)$. This is the simplest and strongest type of convergence behavior for $\sum_{k=1}^{\infty} c_k f(n_k x)$, but it holds only in a few special situations. By Carleson's deep theorem [8], $(\cos 2\pi n x)_{n\geq 1}$ and $(\sin 2\pi n x)_{n\geq 1}$ are convergence systems. More generally, Gaposhkin [16] proved (using Carleson's theorem) the following result:

THEOREM C. *Let $f \in \mathrm{Lip}_\alpha(\mathbf{T})$ for $\alpha > 1/2$ and $\int_{\mathbf{T}} f(t)dt = 0$. Then $\{f(nx)$, $n = 1, 2, \dots\}$ is a convergence system.*

Another classical result, proved by Kac [27] for the Lipschitz class and extended substantially by Gaposhkin [15] is the following

THEOREM D. *Let $f \in L^2(\mathbf{T})$ with $\int_{\mathbf{T}} f(t)dt = 0$ and assume that the square modulus of continuity*

$$\omega_2(\delta, f) = \sup_{0<h\leq\delta} \left\{ \int_0^1 |f(x+h) - f(x)|^2 dx \right\}^{1/2}$$

of f satisfies

$$\omega_2(\delta, f) = \mathcal{O}\Big(\log \frac{1}{\delta}\Big)^{-1-\varepsilon} \qquad (\varepsilon > 0). \tag{2.1}$$

Let (n_k) be an sequence of positive numbers satisfying the Hadamard gap condition

$$n_{k+1}/n_k \geq q > 1 \qquad k = 1, 2, \dots \tag{2.2}$$

Then $f(n_k x)$ is a convergence system.

These theorems describe the known situations when $f(n_k x)$ is a convergence system; note that all conditions of these results are sharp. Gaposhkin [15] showed that Theorem D becomes false if we replace the right-hand side of (2.1) by $\mathcal{O}\big(\log \frac{1}{\delta}\big)^{-1/2}$ and Berkes [4] proved that the the condition $f \in \mathrm{Lip}_\alpha(\mathbf{T})$, $\alpha > 1/2$ in Theorem C and the Hadamard gap condition (2.2) in Theorem D are also best possible: there exists a function $f \in \mathrm{Lip}_{1/2}(\mathbf{T})$ with $\int_{\mathbf{T}} f(t)dt = 0$ such that for any positive sequence (ε_k) tending to 0, there exists an increasing sequence (n_k) of positive integers satisfying

$$n_{k+1}/n_k \geq 1 + \varepsilon_k, \qquad k = 1, 2, \dots \tag{2.3}$$

together with a sequence $\mathbf{c} \in \ell^2$ such that the series $\sum_{k=1}^{\infty} c_k f(n_k x)$ diverges almost everywhere. Going beyond the conditions of Theorems C and D, the almost everywhere convergence behavior of $\sum_{k=1}^{\infty} c_k f(n_k x)$ becomes very complicated and and examples show that the properties of $\sum_{k=1}^{\infty} c_k f(n_k x)$ are determined by a delicate interplay between the coefficient sequence (c_k), the smoothness properties of f and the growth speed and number-theoretic properties of (n_k). In this section we give a detailed study of this behavior and prove several convergence results for such series. Our main interest will be to find convergence criteria of the type $\sum_{k=1}^{\infty} c_k^2 \omega(k) < \infty$ where $\omega(k) \to \infty$ is some positive sequence (called Weyl multiplier) depending on f and (n_k).

Before formulating our results, we first give equivalent reformulations of the convergence system property of $\sum_{k=1}^{\infty} c_k f(n_k x)$ in terms of maximal inequalities. The following result is due to Nikishin [36].

PROPOSITION 2.1. *A pair $(f, \mathcal{N})$ is a convergence system if and only if for any $\varepsilon > 0$, $0 < \delta < 1$ there exist a set $A_{\varepsilon,\delta} \subset (0,1)$ with Lebesgue measure $\geq 1 - \varepsilon$ and a constant $C_{\varepsilon,\delta} > 0$ such that for arbitrary $\mathbf{c} \in \ell^2$ we have*

$$\int_{A_{\varepsilon,\delta}} \sup_{N \geq 1} \left| \sum_{k=1}^{N} c_k f(n_k x) \right|^{1-\delta} dx \leq C_{\varepsilon,\delta} \left(\sum_{k=1}^{\infty} c_k^2 \right)^{(1-\delta)/2}.$$

We now prove an analogous statement involving a weak $(2,2)$ type inequality.

PROPOSITION 2.2. *A pair $(f, \mathcal{N})$ is a convergence system if and only if there exists a constant C such that for any $\mathbf{c} \in \ell^2$ the following maximal inequality holds:*

$$\sup_{t \geq 0} t^2 \lambda \Big\{ \sup_{N \geq 1} \left| S_N^{\mathcal{N}}(\mathbf{c}, f) \right| > t \|\mathbf{c}\|_2 \Big\} \leq C.$$

PROOF. Given a pair $(f, \mathcal{N})$, consider the $L^2(\mathbf{T})$–operators $S_N^{\mathcal{N},f}$, $N = 1, 2 \ldots$ defined via the isomorphism $\mathbf{c} \mapsto g$ if $g \sim \sum_k c_{|k|} e_k$ by

$$S_N^{\mathcal{N},f}(g) = \sum_{k=1}^{N} c_k f(n_k \cdot).$$

A first relevant observation in the proof will concern the following commutation property. Consider the family of pointwise measurable transformations of $\mathbf{T}$ defined for each positive integer j by

$$\tau_j x = jx \bmod 1.$$

For fixed j, the transformation $\tau = \tau_j$ preserves the normalized Lebesgue measure λ (see Halmos [21], p. 5-37), so that τ is an endomorphism of the torus. It has been proved by Philipp [39] (see Theorem 1 on p. 112) that τ is also strongly mixing. Now, let $\mathcal{E}$ denote the family of associated operators on $L^0(\mathbf{T})$:

$$T_j g = g \circ \tau_j.$$

That the T_j's are commuting positive L^2-isometries, preserving 1 is better viewed on Fourier expansion of g, since if $g \sim \sum_{m \in \mathbf{Z}} g_m e_m$, then $T_j f \sim \sum_{m \in \mathbf{Z}} g_m e_{mj}$, which readily implies

$$T_k(T_j g) = T_j(T_k g), \qquad (j, k = 1, 2 \ldots). \tag{2.4}$$

Proceeding next by approximation, we deduce that (2.4) hold for any $g \in L^p(\mathbf{T})$, $0 < p \le \infty$. This in particular implies that the sequence of operators $S_N^{\mathcal{N},f}$ commutes with $\mathcal{E}$: for any $g \in L^2(\mathbf{T})$,

$$S_N^{\mathcal{N},f}(T_j g) = T_j(S_N^{\mathcal{N},f} g), \qquad (N, j = 1, 2 \dots) \tag{2.5}$$

Further, the family $\mathcal{E}$ satisfies a mean ergodic theorem in $L^2(\mathbf{T})$: for any $g \in L^2(\mathbf{T})$,

$$\lim_{J\to\infty} \left\| \frac{1}{J} \sum_{j=1}^{J} T_j g - \int_{\mathbf{T}} g d\lambda \right\|_2 = 0.$$

Since strong convergence implies weak convergence, it follows that for any $u, v \in L^2(\mathbf{T})$ we have

$$\lim_{J\to\infty} \frac{1}{J} \sum_{j=1}^{J} \langle T_j u, v \rangle = \langle u, 1 \rangle \langle v, 1 \rangle.$$

Choosing $u = \chi\{A\}$, $v = \chi\{B\}$ where A, B are Borel sets of $\mathbf{T}$ and χ denotes indicator function, we deduce

$$\lim_{J\to\infty} \frac{1}{J} \sum_{j=1}^{J} \lambda(T_j^{-1} A \cap B) = \lambda(A)\lambda(B).$$

From this it follows easily that for any $a > 1$ and Borel sets A, B of $\mathbf{T}$, there exists $T \in \mathcal{E}$ such that

$$\lambda(T^{-1} A \cap B) \le a\lambda(A)\lambda(B). \tag{2.6}$$

Now Proposition 2.2 states, in terms of operators, that for any $g \in L^2(\mathbf{T})$ we have

$$\sup_{t \ge 0} t^2 \lambda\Big\{ \sup_{N \ge 1} |S_N^{\mathcal{N},f} g| > t\|g\|_2 \Big\} \le C.$$

And this is a consequence of the Continuity Principle (Garsia [18]). □

Proposition 2.2 implies that a pair $(f, \mathcal{N})$ is a convergence system only if the maximal operator

$$\sup_{N \ge 1} \left| S_N^{\mathcal{N}}(\mathbf{c}, f) \right|$$

belongs to $L^p(\mathbf{T})$ with $p < 2$. This has a consequence concerning convergence in mean. Say by analogy that a pair $(f, \mathcal{N})$ is an L^p-convergence system if for any $g \in L^2(\mathbf{T})$ the sequence $\{S_N^{\mathcal{N},f} g, N \ge 1\}$ converges in $L^p(\mathbf{T})$.

COROLLARY 2.1. *Assume that the pair $(f, \mathcal{N})$ is a convergence system. Then, it is also an L^p-convergence system for any $p < 2$.*

PROOF. Define $\omega_R = \sup_{N,M \ge R} \left| S_N^{\mathcal{N}}(\mathbf{c}, f) - S_M^{\mathcal{N}}(\mathbf{c}, f) \right|$. By assumption $\lim_{R\to\infty} \omega_R = 0$ a.e. And by the above remark $\omega_1 \in L^p(\mathbf{T})$, $p < 2$. Thus by Fatou's lemma

$$0 = \mathbf{E} \limsup_{N,M\to\infty} \left| S_N^{\mathcal{N}}(\mathbf{c}, f) - S_M^{\mathcal{N}}(\mathbf{c}, f) \right|^p \ge \limsup_{N,M\to\infty} \mathbf{E} \left| S_N^{\mathcal{N}}(\mathbf{c}, f) - S_M^{\mathcal{N}}(\mathbf{c}, f) \right|^p.$$

□

The previous results summarize the basic equivalence of a.e. convergence results and maximal inequalities for $f(n_k x)$. In Theorem 2.6 at the end of this section we will in fact prove a maximal inequality that leads to various a.e. convergence results for $\sum_{k=1}^{\infty} c_k f(n_k x)$, see Corollaries 2.7–2.9. In the rest of the paper our approach to a.e. convergence will be different: we will use a combination of martingale and quasi-orthogonality arguments as our fundamental technical tools.

Theorems C and D show that the convergence properties of $\sum_{k=1}^{\infty} c_k f(n_k x)$ depend sensitively on the smoothness properties of f and we start with a few preliminary remarks concerning smoothness criteria. Let $f \in L^2(\mathbf{T})$ with $\int_{\mathbf{T}} f(t)dt = 0$ have Fourier series

$$f \sim \sum_{k=1}^{\infty} (a_k \cos 2\pi kx + b_k \sin 2\pi kx) \tag{2.7}$$

and let

$$r_f(N) = \sum_{k=N+1}^{\infty} (a_k^2 + b_k^2). \tag{2.8}$$

Given an integer $m \geq 1$, let $[f]_m$ denote the function in $[0,1)$ which takes the constant value $m \int_{k/m}^{(k+1)/m} f(t)dt$ in the interval $[k/m, (k+1)/m)$ $(k=0,1,...,m-1)$. In probabilistic terms, $[f]_m$ is the conditional expectation of f with respect to the σ-field generated by the intervals $[k/m, (k+1)/m)$, $k = 0, 1, \dots m-1$. Let

$$r_f^*(N) = \|f - [f]_N\|. \tag{2.9}$$

(Here, and in the sequel, $\|\cdot\|$ means L^2 norm, while the L^p norm will be denoted by $\|\cdot\|_p$.) The speed of convergence of $r_f^*(N)$ to zero clearly measures the smoothness of f; for example, if f is a Lip (α) function then $r_f^*(N) = O(n^{-\alpha})$. A simple connection between $r_f(N)$ and $r_f^*(N)$ is given by the following lemma, due essentially to Ibragimov [26]. Its proof will be given after the proofs of Theorems 2.1 and 2.2.

LEMMA 2.1. *Let $\lambda > 1$ and $g(t) = f(\lambda t)$. Then we have for any $m \geq \lambda$*

$$\|g - [g]_m\| \leq C\big((m/\lambda)^{-1/2} + r_f((m/\lambda)^{1/3})\big)$$

where C is a positive constant depending only on f.

In particular, for any $N \geq 1$ we have

$$r_f^*(N) \leq C\big(N^{-1/2} + r_f(N^{1/3})\big). \tag{2.10}$$

Thus if $r_f(N) = O(N^{-\alpha})$ for some $0 < \alpha \leq 1$, then $r_f^*(N) = O(N^{-\alpha/3})$.

Turning to the convergence behavior of $\sum c_k f(n_k x)$, we first study the lacunary case, i.e. we assume that (n_k) grows very rapidly. If (n_k) satisfies the Hadamard gap condition (2.2), then by Theorem D the system $f(n_k x)$ is a convergence system under mild smoothness conditions on f. We investigate now the case when (n_k) grows with a sub-exponential speed, i.e. it satisfies the gap condition

$$n_{k+1}/n_k \geq 1 + \varepsilon_k, \qquad k = 1, 2, \dots$$

where ε_k tends to 0. A remarkable result on trigonometric series with sub-Hadamard gaps was proved by Erdős [10], who showed that if (n_k) is a sequence of positive integers satisfying

$$n_{k+1}/n_k \geq 1 + ck^{-\beta}, \qquad k \geq k_0 \tag{2.11}$$

for some $c > 0$, $0 < \beta < 1/2$, then $\sin 2\pi n_k x$ satisfies the central limit theorem, i.e.

$$\lim_{N\to\infty} \lambda\left\{x \in (0,1) : (N/2)^{-1/2} \sum_{k=1}^{N} \sin 2\pi n_k x \le t\right\} = (2\pi)^{-1/2} \int_{-\infty}^{t} e^{-u^2/2} du.$$

Moreover, this result becomes false for $\beta = 1/2$. Thus, under (2.11) with $0 < \beta < 1/2$ the sequence $\sin 2\pi n_k x$ behaves like a sequence of independent random variables, and this is no more valid if $\beta = 1/2$. Our next theorem gives a strong convergence property of series $\sum_{k=1}^\infty c_k f(n_k x)$ under the Erdős gap condition (2.11). Define, for any $\varrho > 0$

$$\tau_{k,\varrho}(\mathbf{c}) = \sup_{L \ge k^{\varrho+1}} \sum_{\ell=L}^{L+[k^\varrho]} |c_\ell|.$$

THEOREM 2.1. *Let $f \in L^\infty(\mathbf{T})$ with $\int_{\mathbf{T}} f(t)dt = 0$ and $r_f(N) = O(N^{-\alpha})$ for some $\alpha > 0$. Let (n_k) be a sequence of positive integers satisfying the gap condition (2.11) with some $0 < \beta < 1/2$, and let $\mathbf{c} \in \ell^2$ with $\tau_{k,\varrho}(\mathbf{c}) = o(1)$ $(k \to \infty)$ for all $0 < \varrho < 1$. Assume that $\sum_{k=1}^\infty c_k f(n_k x)$ and all of its subseries converge in $L^2(\mathbf{T})$ norm. Then $\sum_{k=1}^\infty c_k f(n_k x)$ also converges a.e.*

It seems likely that Theorem 2.1 remains valid without the technical condition $\tau_{k,\varrho}(\mathbf{c}) = o(1)$, but this remains open. This condition is certainly satisfied if $c_k = O(k^{-1/2})$ which, in turn, holds if $\mathbf{c} \in \ell^2$ and (c_k) is monotone.

Note that if X_k are independent r.v.'s then under suitable moment conditions, mean convergence of $\sum_{k=1}^\infty X_k$ implies a.e. convergence of the same series. Theorem 2.1 establishes a similar property for $\sum_{k=1}^\infty c_k f(n_k x)$. Note that the central limit theorem is in general not valid for $f(n_k x)$ under the gap condition (2.11) with $0 < \beta < 1/2$, despite Erdős' theorem mentioned above. (See Kac [29], p. 646.)

COROLLARY 2.2. *Let $f \in L^\infty(\mathbf{T})$ with $\int_{\mathbf{T}} f(t)dt = 0$ and $r_f(N) = O(N^{-\alpha})$ for some $\alpha > 0$. Let (2.7) be the Fourier expansion of f and assume that the Dirichlet series*

$$\sum_{n=1}^\infty a_n n^{-s}, \quad \text{and} \quad \sum_{n=1}^\infty b_n n^{-s} \tag{2.12}$$

are regular and bounded in the half-plane $\Re(s) > 0$. Let (n_k) be a sequence of positive integers satisfying the gap condition (2.11) with some $0 < \beta < 1/2$. Then $\sum_{k=1}^\infty c_k f(n_k x)$ converges a.e. provided $\mathbf{c} \in \ell^2$ and $c_k = O(k^{-1/2})$.

Corollary 2.2 connects the a.e. convergence of lacunary series $\sum_{k=1}^\infty c_k f(n_k x)$ to the classical Wintner theory, showing that the boundedness and regularity of the Dirichlet series (2.12) for $\Re(s) > 0$ implies not only mean, but actually a.e. convergence in the lacunary case. In Chapter 3 we will show that this result is best possible: if the boundedness condition for the Dirichlet series (2.12) is not satisfied, there exists a sequence (n_k) satisfying (2.11) for all $0 < \beta < 1/2$, together with a positive nonincreasing sequence $\mathbf{c} \in \ell^2$ such that $\sum_{k=1}^\infty c_k f(n_k x)$ diverges almost everywhere. On the other hand, if we are interested in the a.e. convergence of $\sum_{k=1}^\infty c_k f(n_k x)$ under more stringent coefficient conditions like $\sum_{k=1}^\infty c_k^2 \omega(k) < \infty$, $\omega(k) \to \infty$, then the condition on the Dirichlet series can be dropped, as the following result shows.

LEMMA 2.2. *Let $f \in \mathrm{Lip}_\alpha(\mathbf{T})$ for some $0<\alpha\le 1$ and assume that $\int_{\mathbf{T}} f(t)dt = 0$. Let (n_k) be an increasing sequence of positive integers and put*

$$\omega(j) := \max\left(\sum_{1\le \ell\le j} (n_\ell/n_j)^\alpha, \sum_{\ell\ge j} (n_j/n_\ell)^\alpha\right).$$

Then

$$\int_{\mathbf{T}} \left(\sum_{k=1}^N c_k f(n_k x)\right)^2 dx \le C \sum_{k=1}^N c_k^2 \omega(k)$$

with some constant C. In particular, if $\sum_{k=1}^\infty c_k^2\omega(k) < \infty$, $\sum_{k=1}^\infty c_k f(n_k x)$ and all of its subseries converge in $L^2(\mathbf{T})$ norm.

In particular, if $n_k = [\exp(k/(\log k)^\tau)]$, $\tau > 0$, then $\omega(j) \sim \mathrm{const}\cdot(\log j)^\tau$ and in the case $n_k = [\exp(k^\eta)]$, $0 < \eta < 1$, then $\omega(j) \sim \mathrm{const}\cdot j^{1-\eta}$.

We supplement Theorem 2.1 with another result reducing the almost everywhere convergence of $\sum_{k=1}^N c_k f(n_k x)$ to mean convergence under an additional assumption on the size of the tail sums $\sum_{k>N} c_k^2$, or, alternatively, under assuming $\sum_{k=1}^\infty c_k^2 \omega(k) < \infty$ for a suitable $\omega(k) \to \infty$.

THEOREM 2.2. *Let $f \in \mathrm{Lip}_\alpha(\mathbf{T})$ for some $0 < \alpha \le 1$ and assume that $\int_{\mathbf{T}} f(t)dt = 0$. Let (n_k) be an increasing sequence of positive integers and $\mathbf{c} \in \ell^2$. Assume that $\sum_{k=1}^\infty c_k f(n_k x)$ converges in L^2 norm and*

$$\lim_{R\to\infty} \left(\sum_{k>R} c_k^2\right)^{1/2} \left(\sum_{k>R} n_k^{-2}\right)^{1/2} \left(\sum_{k\le R} n_k^\alpha\right)^{1/\alpha} = 0. \tag{2.13}$$

Then $\sum_{k=1}^\infty c_k f(n_k x)$ converges almost everywhere.

If the sequence (n_k) satisfies the Hadamard gap condition (2.2), relation (2.13) trivially holds whenever $\mathbf{c} \in \ell^2$. If, on the other hand, (n_k) grows slower than exponentially, condition (2.13) imposes a restriction on the tail sums $\sum_{k>R} c_k^2$, which is very mild if (n_k) grows near exponentially. For example, if $n_k = [e^{k/(\log k)^\tau}]$ for some $\tau > 0$, then (2.13) reduces to

$$\sum_{k>R} c_k^2 = o\big((\log R)^{-\tau(1+2/\alpha)}\big).$$

If $n_k = [e^{k/(\log\log k)^\tau}]$, $\tau > 0$, then (2.8) becomes

$$\sum_{k>R} c_k^2 = o\big((\log\log R)^{-\tau(1+2/\alpha)}\big),$$

and if $n_k = [e^{k^\gamma}]$, $0 < \gamma < 1$ then we get

$$\sum_{k>R} c_k^2 = o\big(R^{-(1-\gamma)(1+2/\alpha)}\big).$$

The latter case corresponds to the Erdős gap condition (2.11), and thus we see that the conditions of Theorem 2.2 are more restrictive than those of Theorem 2.1. On the other hand, in Theorem 2.2 we do not assume regularity conditions like $c_k = O(k^{-1/2})$.

PROOF OF THEOREM 2.2. We follow Kac [27]. Assume $\sum_{k=1}^{\infty} c_k f(n_k x)$ converges in norm to $f^* \in L^2(\mathbf{T})$. For almost all points t_0

$$f^*(t_0) = \lim_{h \to 0} \frac{1}{h} \int_{t_0}^{t_0+h} f^*(u)du. \tag{2.14}$$

Clearly

$$\int_{t_0}^{t_0+h} f^*(u)du = \sum_{k \geq 1} c_k \int_{t_0}^{t_0+h} f(n_k u)du. \tag{2.15}$$

We shall use the following estimate: there exists a constant C such that for any $0 \leq a < b < 1$ and any positive integer k

$$\left| \int_a^b f(n_k u)du \right| \leq C n_k^{-1}. \tag{2.16}$$

Let χ be the characteristic function of the interval $[a, b]$, extended with period 1 to the whole real line and let

$$\chi(x) = \sum_{m \in \mathbf{Z}} a_m e_m(x)$$

be the Fourier expansion of χ. By Parseval's relation,

$$\int_a^b f(n_k u)du = \sum_{m \in \mathbf{Z}} \varphi_m \overline{a_{n_k m}}.$$

Since χ is of bounded variation, we have

$$a_m = \mathcal{O}(1/|m|),$$

(see Zygmund [55], Vol. I, p. 48) and thus we get

$$\left| \int_a^b f(n_k u)du \right| \leq \left[\sum_{m \in \mathbf{Z}} |\varphi_m|^2 \right]^{1/2} \left[\sum_{m \in \mathbf{Z}} |a_{n_k m}|^2 \right]^{1/2} \leq C\|f\|_2 / n_k.$$

Combining (2.15) with (2.16) gives

$$\left| \int_{t_0}^{t_0+h} f^*(u)du - \sum_{k=1}^{R} c_k \int_{t_0}^{t_0+h} f(n_k u)du \right| \leq C \left[\sum_{k>R} c_k^2 \right]^{1/2} \left[\sum_{k>R} \left(\frac{1}{n_k} \right)^2 \right]^{1/2}.$$

Since f belongs to $\mathrm{Lip}_\alpha(\mathbf{T})$,

$$\left| \sum_{k=1}^{R} c_k \int_{t_0}^{t_0+h} [f(n_k u) - f(n_k t_0)] du \right| \leq C|h|^{1+\alpha} \sum_{k=1}^{R} |c_k| n_k^{\alpha}.$$

Therefore

$$\left| \frac{1}{h} \int_{t_0}^{t_0+h} f^*(u)du - \sum_{k=1}^{R} c_k f(n_k t_0) \right|$$

$$\leq C \left\{ |h|^{-1} \left[\sum_{k>R} c_k^2 \right]^{1/2} \left[\sum_{k>R} \left(\frac{1}{n_k} \right)^2 \right]^{1/2} + |h|^{\alpha} \sum_{k=1}^{R} |c_k| n_k^{\alpha} \right\}.$$

Choosing $h = h_R = \left(\sum_{k=1}^R n_k^\alpha\right)^{-1/\alpha}$ and observing that

$$|h_R|^\alpha \sum_{k=1}^R |c_k| n_k^\alpha = \frac{\sum_{k=1}^R |c_k| n_k^\alpha}{\sum_{k=1}^R n_k^\alpha} \to 0,$$

as R tend to infinity since c_k tend to 0 as k tend to infinity, finally shows in view of condition (2.13)

$$\lim_{R\to\infty} \left| \frac{1}{h_R} \int_{t_0}^{t_0+h_R} f^*(u)du - \sum_{k=1}^R c_k f(n_k t_0) \right| = 0.$$

The proof is completed by combining the above result with (2.14). □

PROOF OF LEMMA 2.2. Let $f \in \mathrm{Lip}_\alpha(\mathbf{T})$ and let

$$f \sim \sum_{k=1}^\infty (a_k \cos 2\pi kx + b_k \sin 2\pi kx)$$

be its Fourier expansion. Then we have (see Zygmund [55], Vol I, p. 241)

$$\sum_{\ell=n+1}^\infty (a_\ell^2 + b_\ell^2) \le D n^{-2\alpha}.$$

Let $j \le k$ be fixed positive integers. Using Parseval's relation yields

$$\int_{\mathbf{T}} f(n_j x) f(n_k x) dx = \sum_{r n_j = s n_k} (a_r a_s + b_r b_s).$$

The relation $j \le k$ together with $rn_j = sn_k$ implies that $r \ge n_k/n_j$. Using the inequality $|a_r a_s + b_r b_s| \le (a_r^2 + b_r^2)^{1/2}(a_s^2 + b_s^2)^{1/2}$, the Cauchy–Schwarz inequality and the previous estimate for the tail sums of $a_\ell^2 + b_\ell^2$ we get

$$\left|\int_{\mathbf{T}} f(n_j x) f(n_k x) dx\right| \le \left[\sum_{r \ge n_k/n_j} (a_r^2 + b_r^2)\right]^{1/2} \left[\sum_{s\ge 1} (a_s^2 + b_s^2)\right]^{1/2} \le B\left(\frac{n_j}{n_k}\right)^\alpha.$$

Thus

$$\begin{aligned}
&\left|\int_{\mathbf{T}} \sum_{1\le j<k\le N} c_j c_k f(n_j x) f(n_k x) dx\right| \\
&\le \sum_{1\le j<k\le N} |c_j||c_k| \left(\frac{n_j}{n_k}\right)^\alpha \le \sum_{1\le j<k\le N} \left(\frac{|c_j|^2 + |c_k|^2}{2}\right)\left(\frac{n_j}{n_k}\right)^\alpha \\
&\le \frac{1}{2}\sum_{k=1}^N c_k^2 \sum_{1\le j<k} \left(\frac{n_j}{n_k}\right)^\alpha + \frac{1}{2}\sum_{j=1}^N c_j^2 \sum_{j<k\le N} \left(\frac{n_j}{n_k}\right)^\alpha \le \sum_{k=1}^N c_k^2 \omega(k),
\end{aligned}$$

proving Lemma 2.2. □

PROOF OF THEOREM 2.1. Without loss of generality we can assume $\alpha \le 1$. As a first step, we approximate the functions $f(n_k x)$ by stepfunctions $\varphi_k(x)$ as follows. Let $2^\ell \le n_k < 2^{\ell+1}$, put $m = \left[\ell + 60\alpha^{-1}\log k\right]$ and let $\varphi_k(x) = [f(n_k x)]_{2^m}$. By Lemma 2.1 we have

$$\|f(n_k x) - \varphi_k(x)\| \le C\left(\frac{2^m}{n_k}\right)^{-\alpha/3} \le C2^{-20\log k} \le Ck^{-10}. \tag{2.17}$$

Choose ϱ so that $\frac{1}{2} \vee \frac{\beta}{1-\beta} < \varrho < 1$ (such a ϱ exists since $\beta < 1/2$) and split the sequence of positive integers into consecutive blocks $\Delta_1, \Delta_1', \Delta_2, \Delta_2', \dots$ so that

$$|\Delta_k| = |\Delta_k'| = [k^\varrho].$$

Set

$$T_k = \sum_{\nu\in\Delta_k} c_\nu f(n_\nu x), \qquad D_k = \sum_{\nu\in\Delta_k} c_\nu \varphi_\nu(x).$$

Clearly, each integer in Δ_k exceeds $(k-1)^\varrho \ge (k-1)^{1/2}$ and thus by (2.17)

$$\|T_k - D_k\| \le C \sum_{\nu=(k-1)^{1/2}}^{\infty} \nu^{-10} \le Ck^{-4}. \tag{2.18}$$

Next we show

LEMMA 2.3. *We have*

$$\mathbf{P}\{|\mathbf{E}\,(D_k \mid \mathcal{F}_{k-1})| \ge k^{-2}\} \le Ck^{-2}, \tag{2.19}$$

where $\mathcal{F}_{k-1}$ denotes the σ-field generated by $D_1, \dots, D_{k-1}$.

PROOF. We first show that

$$|\mathbf{E}\,(T_k \mid \mathcal{F}_{k-1})| \le Ck^{-2}. \tag{2.20}$$

To see this, let r and t denote the largest integer of Δ_{k-1} and the smallest integer of Δ_k, respectively. Let $2^\ell \le n_r < 2^{\ell+1}$, $w = \left[\ell + \frac{60}{\alpha}\log r\right]$. From the definition of φ_ν it is clear that every φ_ν, $1 \le \nu \le r$ takes a constant value on each interval of the form

$$A = [i2^{-w}, (i+1)2^{-w}), \quad 0 \le i \le 2^w - 1 \tag{2.21}$$

and thus each set of the σ-field $\mathcal{F}_{k-1}$ can be written as a union of intervals of the form (2.21). Thus to prove (2.20) it suffices to show that

$$|A|^{-1}\left|\int_A T_k\,dx\right| \le Ck^{-2} \tag{2.22}$$

for any A of the form (2.21). Now for the set A in (2.21) we have

$$\begin{aligned} |A|^{-1}\left|\int_A T_k\,dx\right| &= 2^w\left|\int_{i2^{-w}}^{(i+1)2^{-w}} \sum_{\nu\in\Delta_k} c_\nu f(n_\nu x)dx\right| \\ &= \left|\int_i^{i+1} \sum_{\nu\in\Delta_k} c_\nu f(m_\nu t)dt\right| \end{aligned} \tag{2.23}$$

where $m_\nu = 2^{-w} n_\nu$. Using (2.11), $1+x \geq \exp(x/2)$ for $0 \leq x \leq 1$ and the relations $r \sim t \sim Ck^{\varrho+1}$, $t-r \sim k^\varrho$ we get

$$\begin{aligned}\frac{1}{m_t} = \frac{2^w}{n_t} &\leq \frac{2^\ell r^{60/\alpha}}{n_t} \leq r^{60/\alpha}\frac{n_r}{n_t} \\ &\leq r^{60/\alpha}\prod_{\nu=r}^{t-1}\left(1+\frac{1}{\nu^\beta}\right)^{-1} \leq r^{60/\alpha}\left(1+\frac{1}{t^\beta}\right)^{-(t-r)} \\ &\leq r^{60/\alpha}\exp\left(-\frac{t-r}{2t^\beta}\right) \leq Ck^{120/\alpha}\exp(-Ck^\tau) \\ &\leq Ck^{-3},\end{aligned} \tag{2.24}$$

where $\tau = \varrho-(\varrho+1)\beta > 0$ by the choice of ϱ. By the periodicity of f and $\int_0^1 f dx = 0$ we clearly have for any real L and $\lambda \geq 1$

$$\left|\int_L^{L+1} f(\lambda x) dx\right| \leq \frac{2}{\lambda}\int_0^1 |f(x)| dx$$

and thus (2.24) shows that the last expression of (2.23) cannot exceed

$$\sum_{\nu\in\Delta_k} \frac{C|c_\nu|}{m_\nu} \leq C|\Delta_k|\frac{1}{m_t} \leq Ck^{-2}.$$

Hence we proved (2.22) and thus (2.20).

It is now easy to complete the proof of Lemma 2.3. By (2.18) and well-known properties of conditional expectations we have

$$\left\|\mathbf{E}\left(|D_k - T_k|^2 \mid \mathcal{F}_{k-1}\right)\right\|_1 = \mathbf{E}\,|D_k - T_k|^2 \leq Ck^{-8}$$

and thus by the Markov inequality

$$\mathbf{P}\big\{\mathbf{E}\left(|D_k - T_k| \mid \mathcal{F}_{k-1}\right) \geq k^{-2}\big\} \leq \mathbf{P}\big\{\mathbf{E}\left(|D_k - T_k|^2 \mid \mathcal{F}_{k-1}\right) \geq k^{-4}\big\} \leq Ck^{-4}.$$

Together with (2.20) this yields (2.19).

Set $\overline{D}_k = D_k - \mathbf{E}\left(D_k \mid \mathcal{F}_{k-1}\right)$; clearly $(\overline{D}_k, \mathcal{F}_k)$ is a martingale difference sequence and hence orthogonal. Also,

$$\begin{aligned}\|\mathbf{E}(D_k \mid \mathcal{F}_{k-1})\| &\leq \|\mathbf{E}((D_k - T_k) \mid \mathcal{F}_{k-1})\| + \|\mathbf{E}(T_k \mid \mathcal{F}_{k-1})\| \\ &\leq \|D_k - T_k\| + Ck^{-2} \leq C(k^{-4} + k^{-2})\end{aligned} \tag{2.25}$$

by (2.18) and (2.20). By the assumptions of Theorem 2.1, $\sum_{k=1}^\infty T_k$ converges in $L_2(\mathbf{T})$ norm and thus

$$\left\|\sum_{k=m}^n T_k\right\| \to 0 \qquad \text{as } m, n \to \infty.$$

Consequently, using the orthogonality of $\overline{D}_k$, (2.18) and (2.25) we get

$$\begin{aligned}\left(\sum_{k=m}^{n}\mathbf{E}\overline{D}_k^2\right)^{1/2} &= \left\|\sum_{k=m}^{n}\overline{D}_k\right\| \le \left\|\sum_{k=m}^{n}D_k\right\| + \left\|\sum_{k=m}^{n}\mathbf{E}(D_k \mid \mathcal{F}_{k-1})\right\| \\ &\le \left\|\sum_{k=m}^{n}D_k\right\| + C\sum_{k=m}^{n}k^{-2} \le \left\|\sum_{k=m}^{n}T_k\right\| + C'\sum_{k=m}^{n}k^{-2} \longrightarrow 0\end{aligned} \tag{2.26}$$

as $m, n \to \infty$. Thus $\sum_{k=1}^{\infty}\mathbf{E}\overline{D}_k^2 < \infty$ and thus the martingale convergence theorem implies that $\sum_{k=1}^{\infty}\overline{D}_k$ is a.e. convergent. Now $\sum_{k=1}^{\infty}\mathbf{E}\left(D_k \mid \mathcal{F}_{k-1}\right)$ is a.e. convergent by Lemma 2.4 and the Borel–Cantelli lemma, further $\sum_{k=1}^{\infty}(T_k - D_k)$ is a.e. convergent by (2.18) and the Beppo Levi theorem. Thus $\sum_{k=1}^{\infty}T_k$ is a.e. convergent; for the same reason $\sum_{k=1}^{\infty}T_k'$ is also a.e. convergent, where

$$T_k' = \sum_{\nu\in\Delta_k'} c_\nu f(n_\nu x).$$

Hence setting

$$S_N = \sum_{\nu\le N} c_\nu f(n_\nu x), \quad N_k = 2\sum_{i\le k}[i^\varrho]$$

we proved that S_{N_k} is a.e. convergent. To prove the theorem it remains to show that $M_k \to 0$ a.e. where

$$M_k = \max_{N_k \le N < N_{k+1}} |S_N - S_{N_k}|.$$

Let D denote a constant such that $|f| \le D$. Then by using $N_k \sim Ck^{\varrho+1}$, $N_{k+1} - N_k \sim 2k^\varrho$ and $\tau_{k,\varrho}(\mathbf{c}) = o(1)$ we get

$$M_k \le D\sum_{\nu=N_k+1}^{N_{k+1}} |c_\nu| \le C\tau_{k,\varrho}(\mathbf{c}) = o(1).$$

Hence Theorem 2.1 is proved. □

PROOF OF LEMMA 2.1. We follow Berkes [2], p. 337–338. Let

$$f \sim \sum_{k=1}^{\infty}(a_k\cos 2\pi kx + b_k \sin 2\pi kx)$$

be the Fourier expansion of f and write $f = f_1 + f_2$ where

$$f_1 = \sum_{k=1}^{N}(a_k\cos 2\pi kx + b_k \sin 2\pi kx), \quad f_2 = f - f_1,$$

N is an integer to be specified later. If $g(x) = f(\lambda x)$ then we have $g = g_1 + g_2$, where $g_1(x) = f_1(\lambda x)$, $g_2(x) = f_2(\lambda x)$. Evidently

$$|\cos\beta x - [\cos\beta x]_m| \le \beta/m, \quad |\sin\beta x - [\sin\beta x]_m| \le \beta/m$$

for any $\beta > 0$ and thus using

$$g_1(x) = \sum_{k=1}^{N}(a_k\cos 2\pi k\lambda x + b_k \sin 2\pi k\lambda x)$$

and the linearity of the operation $g \longrightarrow [g]$ and the fact that $\|[g]_m\| \le \|g\|$ we get

$$\begin{aligned}|g_1 - [g_1]_m| &\le \sum_{k=1}^{N} 2\pi k\lambda(|a_k| + |b_k|)m^{-1} \\ &\le 2\pi\lambda m^{-1}\left(\sum_{k=1}^{N} k^2\right)^{1/2}\left[\left(\sum_{k=1}^{\infty} a_k^2\right)^{1/2} + \left(\sum_{k=1}^{\infty} b_k^2\right)^{1/2}\right] \le C\lambda m^{-1}N^{3/2}\end{aligned} \tag{2.27}$$

with some constant C depending on f. Further, by the periodicity of f and f_1 we have

$$\begin{aligned}\|g_2 - [g_2]_m\|^2 &\le 4\|g_2\|^2 = 4\int_0^1 f_2(\lambda x)^2 dx = 4\lambda^{-1}\int_0^\lambda f_2(t)^2 dt \\ &\le 4\lambda^{-1}\int_0^{[\lambda]+1} f_2(t)^2 dt \\ &= 4\lambda^{-1}([\lambda]+1)\int_0^1 f_2(t)^2 dt \le 8\|f - f_1\|^2 = 8r_f(N).\end{aligned} \tag{2.28}$$

Using relations (2.27)–(2.28) we get

$$\|g - [g]_m\| \le C(\lambda m^{-1}N^{3/2} + r(N)) \tag{2.29}$$

whence the statement of the lemma follows by choosing $N = [(m/\lambda)^{1/3}]$.

We turn now to the nonlacunary case, i.e. the case when no growth condition on (n_k) is assumed. As we already indicated, in this case the number-theoretic structure of the sequence (n_k) will play an important role in the convergence behavior of $\sum_{k=1}^\infty c_k f(n_k x)$. Before formulating our results, we first recall a useful notion from the theory of orthogonal series. Let (f_n) be a sequence of functions belonging to $L^2(0,1)$ and let $a_{j,k} = \int_0^1 f_j(x) f_k(x)dx$. We call (f_n) *quasi-orthogonal* when the quadratic form defined on ℓ^2 by $(x_n)_n \mapsto \sum_{h,k} a_{h,k}x_h x_k$ is bounded. The important consequence of this property is that for any sequence $\mathbf{c} = \{c_n, n \ge 1\} \in \ell^2$, the series $\sum_{n=1}^\infty c_n f_n$ converges in $L^2(0,1)$ norm.

A sequence $\mathbf{c} = \{c_n, n \ge 1\} \in \ell^2$ will be said *universal* if the series $\sum c_n \psi_n$ converges almost everywhere for every orthonormal system of functions $(\psi_n)_n$. By a theorem of Schur (see e.g. Olevskii [38], p. 56) if $\mathbf{c}$ is universal, then the series $\sum c_n f_n$ converges almost everywhere for any quasi-orthogonal system of functions (f_n). Typical examples of universal sequences are those produced by the general results from the theory of orthogonal series, like the Rademacher–Mensov theorem, implying that a sequence (c_n) is universal if $\sum_{k=1}^\infty c_k^2(\log k)^2 < \infty$. The notion of quasi-orthogonal systems is therefore of particular relevance in the study of the convergence in mean and/or almost everywhere of series $\sum_n c_n f(n_k x)$. In this direction, we will establish the following general result. Here, and in the sequel, let $L(x) = \log(x \vee 1)$ for $x \in \mathbf{R}$.

THEOREM 2.3. *Let $f \in L^2(\mathbf{T})$ with $\int_{\mathbf{T}} f(x)dx = 0$. Let (n_k) be an increasing sequence of positive integers and assume that there exists a sequence (C_k) of positive integers such that*

$$\sum_{k=1}^{\infty} r_f^*(C_k)^2 < \infty \tag{2.30}$$

and

$$\sup_{h\geq 1} C_h \sum_{k>h} \frac{(n_h, n_k)}{n_k} L\left(\frac{(n_h, n_k)C_k}{n_h}\right) < \infty. \tag{2.31}$$

Then the series $\sum_{k=1}^{\infty} c_k f(n_k x)$ *converges a.e. for any universal sequence* $\mathbf{c}$, *in particular if* $\sum_{k=1}^{\infty} c_k^2 (\log k)^2 < \infty$.

The following theorem describes what happens if condition (2.31) of Theorem 2.3 is not assumed.

THEOREM 2.4. *Let* $f \in L^2(\mathbf{T})$ *with* $\int_{\mathbf{T}} f(x)dx = 0$. *Let* (n_k) *be an increasing sequence of positive integers and assume that there exists a sequence* (C_k) *of positive integers and a positive nondecreasing sequence* (λ_k) *such that* $\lambda_{2k}/\lambda_k = O(1)$ *and*

$$\sum_{k=1}^{\infty} r_f^*(C_k)^2/\lambda_k < \infty \tag{2.32}$$

$$\sup_{1\leq h\leq N} C_h \sum_{h\leq k\leq N} \frac{(n_h, n_k)}{n_k} L\left(\frac{(n_h, n_k)C_k}{n_h}\right) \leq \lambda_N. \tag{2.33}$$

Then $\sum_{k=1}^{\infty} c_k f(n_k x)$ *converges a.e. provided* $\sum_{k=1}^{\infty} c_k^2 (\log k)^2 \lambda_k < \infty$.

Choosing the sequences (C_k) and (λ_k) optimally in Theorem 2.4 requires a "balancing" act, but giving up a little accuracy, such sequences are easy to find: first choose (C_k) so that (2.32) holds with $\lambda_k = 1$ and then choose λ_k so that (2.33) holds.

The following analogue of Theorem 2.4 is easier to formulate and prove, but still have useful applications.

THEOREM 2.5. *Let* $f \in L^2(\mathbf{T})$ *with* $\int_{\mathbf{T}} f(x)dx = 0$ *and with Fourier coefficients satisfying* $a_k = O(k^{-\alpha})$, $b_k = O(k^{-\alpha})$, $\alpha > 1/2$. *Let* (n_k) *be an increasing sequence of positive integers and let* (λ_k) *be a positive nondecreasing sequence such that* $\lambda_{2k}/\lambda_k = O(1)$ *and*

$$\sup_{1\leq h\leq N} \sum_{k=1}^{N} \langle n_h, n_k \rangle^{\alpha} \leq \lambda_N.$$

Then $\sum_{k=1}^{\infty} c_k f(n_k x)$ *converges a.e. provided* $\sum_{k=1}^{\infty} c_k^2 (\log k)^2 \lambda_k < \infty$.

It is worth comparing Theorem 2.5 with Theorem 1.1 giving mean convergence criteria for $\sum_{k=1}^{\infty} c_k f(n_k x)$ under similar, but weaker conditions.

Before proving Theorems 2.3–2.5, we give some applications. They should be compared with Corollaries 1.1-1.3 concerning mean convergence.

COROLLARY 2.3. *Let* $f \in L^2(\mathbf{T})$ *with* $\int_{\mathbf{T}} f(x)dx = 0$ *and* $r_f^*(n) = O(n^{-\alpha})$. *Let* (n_k) *be an increasing sequence of coprime integers such that* $n_k \geq k^{\beta}$ *with some* $\beta > 1 + 1/(2\alpha)$. *Then* $\sum_{k=1}^{\infty} c_k f(n_k x)$ *converges a.e. for any universal* $\mathbf{c}$.

COROLLARY 2.3*. *Let* $f \in L^2(\mathbf{T})$ *with* $\int_{\mathbf{T}} f(x)dx = 0$ *and with Fourier coefficients satisfying* $a_k = O(k^{-\alpha})$, $b_k = O(k^{-\alpha})$, $\alpha > 1/2$. *Let* (n_k) *be an increasing sequence of pairwise coprime integers such that* $\sum_{k=1}^{\infty} n_k^{-\alpha} < \infty$. *Then* $\sum_{k=1}^{\infty} c_k f(n_k x)$ *converges a.e. for any universal* $\mathbf{c}$.

The assumptions of Corollaries 2.3 and 2.3* on f are different, but the conclusions are the same.

COROLLARY 2.4. *Let $f \in L_2(\mathbf{T})$ have Fourier-coefficients $O(1/k)$ (for example, let $f \in BV(\mathbf{T})$) with $\int_{\mathbf{T}} f(x)dx = 0$ and let (n_k) be a sequence of integers such that for any $d \geq 1$ we have $\sum_{d|n_k} n_k^{-1} \leq A/d$ with an absolute constant A. Then $\sum_{k=1}^{\infty} c_k f(n_k x)$ converges a.e. provided*

$$\sum_{k=1}^{\infty} c_k^2 (\log k)^2 \log n_k < \infty.$$

COROLLARY 2.5. *Let $f \in L^2(\mathbf{T})$ with $\int_{\mathbf{T}} f(x)dx = 0$ and $r_f^*(n) = O(n^{-\alpha})$. Then the series $\sum_{k=1}^{\infty} c_k f(kx)$ converges a.e. provided*

$$\sum_{k=1}^{\infty} c_k^2 k^{\beta} < \infty \qquad \text{for some } \beta > 1/(1+2\alpha).$$

COROLLARY 2.5*. *Let $f \in L^2(\mathbf{T})$ with $\int_{\mathbf{T}} f(x)dx = 0$ and with Fourier coefficients satisfying $a_k = O(k^{-\alpha})$, $b_k = O(k^{-\alpha})$, $1/2 < \alpha < 1$. Then $\sum_{k=1}^{\infty} c_k f(kx)$ converges a.e. provided*

$$\sum_{k=1}^{\infty} c_k^2 k^{1-\alpha} (\log k)^2 < \infty.$$

COROLLARY 2.6. *Let $f \in L_2(\mathbf{T})$ with $\int_{\mathbf{T}} f(x)dx = 0$ and with Fourier coefficients satisfying $a_k = O(k^{-\alpha})$, $b_k = O(k^{-\alpha})$, $\alpha > 1/2$. Let $n_k = k^r$, where r is an integer with $r \geq 2$. Then $\sum_{k=1}^{\infty} c_k f(n_k x)$ converges a.e. provided*

$$\sum_{k=1}^{\infty} c_k^2 (\log k)^2 < \infty.$$

PROOF OF THEOREM 2.5. We follow the proof of Theorem 1.1 with minor modifications, using the same notations. The assumption $\sum_{k=1}^{\infty} c_k^2 (\log k)^2 \lambda_k < \infty$ and the estimates in the second line of (1.18) with $\gamma = 2$ show that $\sum_{k=1}^{\infty} \nu^2 \int_{\mathbf{T}} Z_\nu^2 dt < \infty$ and thus $\sum_{k=1}^{\infty} \nu^2 Z_\nu^2 < \infty$ almost everywhere. Hence (1.17) implies that $\sum_{k=2^m+1}^{2^n} f(n_k t) \to 0$ almost everywhere as $m, n \to \infty$ and thus the partial sums $\sum_{k=1}^{2^N} c_k f(n_k x)$ converge a.e. Now (1.16), the monotonicity of λ_n, $\lambda_{2n}/\lambda_n = O(1)$ and the Rademacher–Mensov inequality (see e.g. Zygmund [55], Vol. II, p. 193) imply

$$\begin{aligned} &\int_{\mathbf{T}} \max_{2^N+1 \leq m \leq 2^{N+1}} \left(\sum_{k=2^N+1}^{m} c_k f(n_k t) \right)^2 dt \\ &\leq C_3 \lambda_{2^{N+1}} \left(\sum_{k=2^N+1}^{2^{N+1}} c_k^2 \right) (\log 2^N)^2 \leq C_4 \sum_{k=2^N+1}^{2^{N+1}} c_k^2 (\log k)^2 \lambda_k \end{aligned} \tag{2.34}$$

Summing these relations for $N = 1, 2, \ldots$ and using $\sum_{k=1}^{\infty} c_k^2 (\log k)^2 \lambda_k < \infty$, it follows that

$$\max_{2^N+1 \leq m \leq 2^{N+1}} \left(\sum_{k=2^N+1}^{m} c_k f(n_k t) \right)^2 \to 0 \qquad \text{a.e.}$$

completing the proof of Theorem 2.5. $\square$

PROOF OF THEOREM 2.4. Let $f_k = [f]_{C_k}(n_k\cdot)$. By the Cauchy–Schwarz inequality we get

$$\sum_{k=1}^{\infty} |c_k| \|f(n_k\cdot) - f_k(\cdot)\| = \sum_{k=1}^{\infty} |c_k| \|f(\cdot) - [f]_{C_k}(\cdot)\| = \sum_{k=1}^{\infty} |c_k| r_f^*(C_k)$$
$$\leq \left(\sum_{k=1}^{\infty} c_k^2 \lambda_k\right)^{1/2} \left(\sum_{k=1}^{\infty} r_f^*(C_k)^2/\lambda_k\right)^{1/2} < \infty$$

by the assumptions of Theorem 2.4. It follows that $\sum_{k=1}^{\infty} |c_k| |f(n_k t) - f_k(t)\|$ converges a.e. and thus the series $\sum_{k=1}^{\infty} c_k f(n_k t)$ converges almost everywhere if and only if the series $\sum_{k=1}^{\infty} c_k f_k(t)$ does. The problem thus reduces to the study of the last series, and to do this, we will analyze the correlation properties of the functions f_k. Define, for any non-empty interval $\dot\pi$ of $\mathbf{T}$,

$$f^{\dot\pi} = \frac{1}{|\dot\pi|} \int_{\dot\pi} f(u)du. \tag{2.35}$$

Then

$$[f]_{C_n}(x) = \sum_{\dot\pi \in \Pi_n} f^{\dot\pi} \chi(\dot\pi)(x),$$

where Π_n denotes the partition of $[0, 1)$ defined by the subdivision

$$[(j-1)/C_n, j/C_n) \qquad j = 1, \dots, C_n. \tag{2.36}$$

Since $\int_{\mathbf{T}} f(f)dt = 0$, we have

$$[f]_{C_n}(x) = \sum_{\dot\pi \in \Pi_n} f^{\dot\pi} \left[\chi(\dot\pi)(x) - |\dot\pi|\right] \tag{2.37}$$

and consequently for $h \leq k$ we get

$$\langle f_h, f_k \rangle = \sum_{\dot\pi \in \Pi_h} \sum_{\dot\pi' \in \Pi_k} f^{\dot\pi} f^{\dot\pi'} \langle \chi_{\dot\pi}(\{n_h y\}) - |\dot\pi|, \chi_{\dot\pi'}(\{n_k y\}) - |\dot\pi'| \rangle, \tag{2.38}$$

where the indicators are extended with period 1. Here $\langle f, g \rangle = \int_0^1 f(t)g(t)dt$ denotes the correlation of the functions f, g, in contrast to the notation $\langle m, n \rangle = (m, n)/[m, n]$ $(m, n \in \mathbb{N})$ used previously. However, as correlation will be used solely in this proof, this will not lead to confusion. We thus see that the calculation of $\langle f_h, f_k \rangle$ reduces to estimating the correlation for indicators of intervals. Let $0 \leq a < b < 1$. It is classical to expand the indicator function $\chi([a, b))(x)$ in a Fourier series, and one gets

$$\chi([a, b))(x) = b - a + \sum_{n \in \mathbf{Z}^*} \left(\frac{-1}{2i\pi n}\right) \{e^{-2i\pi nb} - e^{-2i\pi na}\} e^{2i\pi nx}$$
$$= b - a \tag{2.39}$$
$$+ \sum_{n=1}^{\infty} \frac{1}{\pi n} \{\sin 2\pi nx (\cos 2\pi nb - \cos 2\pi na) + \cos 2\pi nx (\sin 2\pi nb - \sin 2\pi na)\},$$

for almost all x. Now, let $0 \le a < b < c < d < 1$. Put $\varphi = \chi([a,b))$, $\psi = \chi([c,d))$, and $\bar\varphi = \varphi - (b-a)$, $\bar\psi = \psi - (d-c)$. We study for given positive integers h and k the correlation of the functions $\bar\varphi_h = \bar\varphi(hx)$, $\bar\psi_k = \bar\psi(kx)$. Put for $u, v \in \mathbf{T}$ and integer n,

$$\delta_n(u,v) = e^{-2i\pi nv} - e^{-2i\pi nu}.$$

Then,

$$\begin{aligned} \bar\varphi(hx) &= \sum_{n\in\mathbf{Z}^*}\left(\frac{-1}{2i\pi n}\right)e^{2i\pi nhx}\delta_n(a,b), \\ \bar\psi(kx) &= \sum_{m\in\mathbf{Z}^*}\left(\frac{-1}{2i\pi m}\right)e^{2i\pi mkx}\delta_m(c,d), \end{aligned} \tag{2.40}$$

so that

$$\begin{aligned} \langle\bar\varphi_h,\bar\psi_k\rangle &= \sum_{n\in\mathbf{Z}^*}\sum_{m\in\mathbf{Z}^*}\frac{1}{4\pi^2 mn}\delta_n(a,b)\delta_{-m}(c,d)\int_{\mathbf{T}} e^{2i\pi(nh-mk)x}dx \\ &= \sum_{\substack{m,n\in\mathbf{Z}^* \\ nh-mk=0}}\frac{1}{4\pi^2 mn}\delta_n(a,b)\delta_{-m}(c,d). \end{aligned} \tag{2.41}$$

The equation $nh - mk = 0$ has solutions given by $n = \mu k/(h,k)$ and $m = \mu h/(h,k)$, $\mu = 1, 2\ldots$. Thus,

$$\begin{aligned} &\langle\bar\varphi_h,\bar\psi_k\rangle \\ &= \frac{\langle h,k\rangle}{4\pi^2}\sum_{\mu=1}^{\infty}\frac{1}{\mu^2}\big\{\delta_{\mu k/(h,k)}(a,b)\delta_{-\mu h/(h,k)}(c,d) + \delta_{-\mu k/(h,k)}(a,b)\delta_{\mu h/(h,k)}(c,d)\big\}. \end{aligned}$$

It remains to compute $\delta_n(a,b)\delta_{-m}(c,d) + \delta_{-n}(a,b)\delta_m(c,d)$. But, a simple calculation shows

$$\begin{aligned} &\delta_n(a,b)\delta_{-m}(c,d) + \delta_{-n}(a,b)\delta_m(c,d) \\ &= 2\big\{\cos 2\pi(nb-md) - \cos 2\pi(nb-mc) - \cos 2\pi(na-md) + \cos 2\pi(na-mc)\big\} \\ &= 2\sin 2\pi m(d-c)\big\{\sin 2\pi(2nb - m(c+d)) - \sin 2\pi(2na - m(c+d))\big\} \\ &= 4\sin 2\pi m(d-c)\sin 2\pi n(b-a)\cos 2\pi(n(a+b) - m(c+d)). \end{aligned}$$

Therefore,

$$\begin{aligned} &\langle\bar\varphi_h,\bar\psi_k\rangle \\ &= \frac{\langle h,k\rangle}{\pi^2}\sum_{\mu=1}^{\infty}\frac{1}{\mu^2}\sin 2\pi\frac{\mu h(d-c)}{(h,k)}\sin 2\pi\frac{\mu k(b-a)}{(h,k)}\cos 2\pi\frac{\mu(k(a+b) - h(c+d))}{(h,k)}. \end{aligned}$$

It follows that

$$\begin{aligned} \big|\langle\bar\varphi_h,\bar\psi_k\rangle\big| &\le \frac{\langle h,k\rangle}{\pi^2}\sum_{\mu=1}^{\infty}\frac{1}{\mu^2}\left|\sin 2\pi\frac{\mu h(d-c)}{(h,k)}\right|\left|\sin 2\pi\frac{\mu k(b-a)}{(h,k)}\right| \\ &\le \frac{2}{\pi}\min\left\{\sum_{\mu=1}^{\infty}\frac{1}{\mu^2}\left(\frac{\mu h(d-c)}{[h,k]}\wedge\langle h,k\rangle\right), \sum_{\mu=1}^{\infty}\frac{1}{\mu^2}\left(\frac{\mu k(b-a)}{[h,k]}\wedge\langle h,k\rangle\right)\right\}. \end{aligned} \tag{2.42}$$

Now, if $\frac{(h,k)}{h(d-c)} > 1$

$$\sum_{\mu \le \frac{(h,k)}{h(d-c)}} \frac{1}{\mu^2}\Big(\frac{\mu h(d-c)}{[h,k]} \wedge \langle h,k\rangle\Big) \le \frac{h(d-c)}{[h,k]} \sum_{\mu \le \frac{(h,k)}{h(d-c)}} \frac{1}{\mu} \le C\frac{h(d-c)}{[h,k]} \log \frac{(h,k)}{h(d-c)},$$

and,

$$\sum_{\mu > \frac{(h,k)}{h(d-c)}} \frac{1}{\mu^2}\Big(\frac{\mu h(d-c)}{[h,k]} \wedge \langle h,k\rangle\Big) \le \langle h,k\rangle \sum_{\mu > \frac{(h,k)}{h(d-c)}\ \frac{1}{\mu^2} \le C\langle h,k\rangle \frac{h(d-c)}{(h,k)}} = C\frac{h(d-c)}{[h,k]}.$$

Thus

$$\sum_{\mu=1}^{\infty} \frac{1}{\mu^2}\Big(\frac{\mu h(d-c)}{[h,k]} \wedge \langle h,k\rangle\Big) \le C\frac{h(d-c)}{[h,k]} \log \frac{(h,k)}{h(d-c)}. \tag{2.43a}$$

If $\frac{(h,k)}{h(d-c)} \le 1$, then $1 \le \frac{h(d-c)}{(h,k)}$ and

$$\sum_{\mu=1}^{\infty} \frac{1}{\mu^2}\Big(\frac{\mu h(d-c)}{[h,k]} \wedge \langle h,k\rangle\Big) \le C\langle h,k\rangle \le C\langle h,k\rangle \frac{h(d-c)}{(h,k)} = C\frac{h(d-c)}{[h,k]}. \tag{2.43b}$$

In both cases we get

$$\sum_{\mu=1}^{\infty} \frac{1}{\mu^2}\Big(\frac{\mu h(d-c)}{[h,k]} \wedge \langle h,k\rangle\Big) \le C\frac{h(d-c)}{[h,k]} L\Big(\frac{(h,k)}{h(d-c)}\Big). \tag{2.44}$$

Therefore,

$$\begin{aligned} &\big|\langle \bar\varphi_h, \bar\psi_k\rangle\big| \\ &\le C \min\Big\{\frac{h(d-c)}{[h,k]} L\Big(\frac{(h,k)}{h(d-c)}\Big), \frac{k(b-a)}{[h,k]} L\Big(\frac{(h,k)}{k(b-a)}\Big), \langle h,k\rangle\Big\}. \end{aligned} \tag{2.45}$$

Return now to (2.38). We deduce from (2.45)

$$\big|\langle \chi_{\dot\pi}(n_h y) - |\dot\pi|, \chi_{\dot\pi'}(n_k y) - |\dot\pi'|\rangle\big| \le C\frac{n_h|\dot\pi'|}{[n_h,n_k]} L\Big(\frac{(n_h,n_k)}{n_h|\dot\pi'|}\Big),$$

so that

$$\begin{aligned} \big|\langle f_h, f_k\rangle\big| &\le C \sum_{\dot\pi\in\Pi_h} |f^{\dot\pi}| \sum_{\dot\pi'\in\Pi_k} |f^{\dot\pi'}||\dot\pi'| \frac{n_h}{[n_h,n_k]} L\Big(\frac{(n_h,n_k)}{n_h|\dot\pi'|}\Big) \\ &\le C \sum_{\dot\pi\in\Pi_h} |f^{\dot\pi}| \sum_{\dot\pi'\in\Pi_k} \Big|\int_{\dot\pi'} f(u)du\Big| \frac{n_h}{[n_h,n_k]} L\Big(\frac{(n_h,n_k)C_{N_k}}{n_h}\Big) \\ &\le C\|f\|_1 \sum_{\dot\pi\in\Pi_h} |f^{\dot\pi}| \Big(\frac{|\dot\pi|}{|\dot\pi|}\Big) \frac{n_h}{[n_h,n_k]} L\Big(\frac{(n_h,n_k)C_k}{n_h}\Big) \\ &\le C\|f\|_1^2 \frac{n_h C_h}{[n_h,n_k]} L\Big(\frac{(n_h,n_k)C_k}{n_h}\Big). \end{aligned} \tag{2.46}$$

Therefore, for $h \le k$

$$\big|\langle f_h, f_k\rangle\big| \le C\|f\|_1^2 \frac{(n_h,n_k)C_h}{n_k} L\Big(\frac{(n_h,n_k)C_k}{n_h}\Big). \tag{2.47}$$

Thus using (2.33) we get

$$\int_{\mathbf{T}}\Big(\sum_{k=1}^{N} c_k f_k\Big)^2 dx = \Big|\sum_{h,k=1}^{N}\langle f_h, f_k\rangle c_h c_k\Big| \le \sum_{h,k=1}^{N}\langle f_h, f_k\rangle \frac{1}{2}(c_h^2+c_k^2) \le \Big(\sum_{k=1}^{N} c_k^2\Big)\lambda_N$$

which corresponds to relation (1.16) in the proof of Theorem 1.1. The argument is now completed by following the proof of Theorem 2.5.

Proof of Theorem 2.3. This is a special case of the previous proof for $\lambda_n = O(1)$.

Proof of Corollaries 2.3, 2.3*. Since in Corollary 2.3 we have $\beta > 1 + 1/(2\alpha)$, we can choose $\gamma > 0$ such that $2\alpha\gamma > 1$ and $\beta > \gamma + 1$. Let $C_k = k^\gamma$, then (2.30) is trivially satisfied and the expression after the sup in (2.31) is at most

$$h^\gamma \sum_{k>h} \frac{1}{n_k} L(C_k) \le K h^\gamma \sum_{k>h} k^{-\beta}\log k = O(h^{(\gamma-\beta+1)}\log h) = O(1).$$

for some constant K. Hence Corollary 2.3 follows from Theorem 2.3. A similar calculation shows that Corollary 2.3* follows from Theorem 2.5.

Proof of Corollary 2.4. This is immediate from Theorem 2.5 and estimate (1.22) in Section 1. □

Proof of Corollary 2.5. Let $C_n = n^\gamma$ where γ will be determined later. Observe that

$$C_h \sum_{k=h}^{n} \frac{(h,k)}{k} L\left(\frac{(h,k)C_k}{h}\right) \le \log C_n \sum_{k=h}^{n} \frac{(h,k)C_h}{k}. \tag{2.48}$$

Fix $1 \le h \le N$ and $d|h$ and compute the last sum in (2.48) for those $h \le k \le n$ such that $(h,k) = d$. This restricted sum clearly cannot exceed

$$C_h \sum_{1\le k\le n, d|k} \frac{d}{k} \le C_n \sum_{l=1}^{[n/d]} \frac{1}{l} \le C_n \log n. \tag{2.49}$$

Now summing for all $d|h$, we have to multiply the result with the number of divisors of h, which is known to be at most $A(\varepsilon)h^\varepsilon \le A(\varepsilon)n^\varepsilon$ for any $\varepsilon > 0$, and thus the first sum in (2.48) is at most $A(\varepsilon)n^\varepsilon C_n \log C_n \log n = O(n^{\gamma+2\varepsilon})$. Thus choosing $\lambda_n = n^{\gamma+2\varepsilon}$, condition (2.33) of Theorem 2.4 is satisfied. Now $r_f^*(C_k) = O(C_k^{-\alpha}) = O(k^{-\gamma\alpha})$ and thus (2.32) will hold if $\gamma = 1/(1+2\alpha)$. As ε can be chosen arbitrarily small, Corollary 2.5 follows from Theorem 2.4. □

Proof of Corollary 2.5*. This is immediate from Theorem 2.5 and the last statement of Lemma 1.1. □

Proof of Corollary 2.6. Let $n_k = k^r$ for some integer $r \ge 2$. Clearly $\langle n_h, n_k\rangle = \langle h, k\rangle^r$ and thus Corollary 2.6 follows from Theorem 2.5 and the first statement of Lemma 1.1. □

To conclude this chapter, we prove a maximal inequality providing a further way to prove a.e. convergence results for $\sum_{k=1}^\infty c_k f(n_k x)$.

THEOREM 2.6. *Let $f \in L_2(\mathbf{T})$ with $\int_{\mathbf{T}} f(t)dt = 0$ have Fourier expansion*

$$f \sim \sum_{k=1}^{\infty} (a_k \cos 2\pi kx + b_k \sin 2\pi kx)$$

and put

$$S_N(x) = \sum_{k \le N} c_k f(kx).$$

Then for any nondecreasing sequence $m_k \to \infty$ of positive integers we have

$$\int_0^1 \max_{M \le N} |S_M(x)| dx \le \sum_{k \le N} |c_k| r_f(m_k)^{1/2} + A \sum_{l=1}^{m_N} (|a_l| + |b_l|) \left(\sum_{k=d_l}^{N} c_k^2 \right)^{1/2} \tag{2.50}$$

where $d_l = \min\{k : m_k \ge l\}$ is the inverse function of m_k and A is an absolute constant.

If the Fourier-series of f is absolutely convergent, i.e. $\sum_{l=1}^{\infty}(|a_l| + |b_l|) < \infty$, then choosing m_k so large that $\sum_{k=1}^{\infty} r_f(m_k) < \infty$, the right-hand side of (2.50) is at most $C(\sum_{k=1}^{N} c_k^2)^{1/2}$, and thus the statement lemma reduces to

$$\int_0^1 \max_{M \le N} |S_M(x)| dx \le C \left(\sum_{k=1}^{N} c_k^2 \right)^{1/2} \tag{2.51}$$

which is an extension of Hunt's inequality (see [25]). (Actually, the proof of Theorem 2.6 uses Hunt's inequality.) In particular, it follows that if the Fourier-series of f is absolutely convergent (for example, if f belongs to the Lip (α) class, $\alpha > 1/2$), then $\sum_{k=1}^{\infty} c_k f(kx)$ converges a.e. provided $\mathbf{c} \in \ell^2$. This result is due to Gaposhkin [16]. In contrast to Theorem 2.3, Theorem 2.6 loses the number-theoretic connection, but in the case $n_k = k$ it leads, despite the simplicity of its proof, to sharper results than the quasi-orthogonality method of Theorem 2.3, as the applications below will show.

PROOF OF THEOREM 2.6. For simplicity we assume that the Fourier-expansion of f is a pure cosine series (i.e. $b_l = 0$); the general case can be treated similarly. Let $k \ge 1$ and write $f = f_k + g_k$ where

$$f_k(x) = \sum_{l=1}^{m_k} a_l \cos 2\pi lx, \qquad g_k(x) = \sum_{l=m_k+1}^{\infty} a_l \cos 2\pi lx,$$

then

$$S_N(x) = T_N^{(1)} + T_N^{(2)}$$

where

$$T_N^{(1)} = \sum_{k \le N} c_k f_k(kx), \qquad T_N^{(2)} = \sum_{k \le N} c_k g_k(kx).$$

Clearly

$$|T_N^{(2)}| \le \sum_{k \le N} |c_k| |g_k(kx)|$$

and thus

$$\max_{M \le N} |T_M^{(2)}| \le \sum_{k \le N} |c_k| |g_k(kx)|.$$

Hence

$$\int_0^1 \max_{M\le N} |T_M^{(2)}|dx \le \sum_{k\le N} |c_k| \|g_k(kx)\|_1 \le \sum_{k\le N} |c_k| r_f(m_k)^{1/2}. \tag{2.52}$$

On the other hand,

$$\begin{aligned} |T_N^{(1)}| &= \left| \sum_{k\le N} c_k \sum_{l=1}^{m_k} a_l \cos 2\pi klx \right| \\ &= \left| \sum_{l=1}^{m_N} a_l \sum_{k=d_l}^{N} c_k \cos 2\pi klx \right| \le \sum_{l=1}^{m_N} |a_l| \left| \sum_{k=d_l}^{N} c_k \cos 2\pi klx \right|. \end{aligned}$$

Thus

$$\max_{M\le N} |T_M^{(1)}| \le \sum_{l=1}^{m_N} |a_l| \max_{M\le N} \left| \sum_{k=d_l}^{M} c_k \cos 2\pi klx \right|$$

and thus using Hunt's inequality we get

$$\int_0^1 \max_{M\le N} |T_M^{(1)}|dx \le A \sum_{l=1}^{m_N} |a_l| \left(\sum_{k=d_l}^{N} c_k^2 \right)^{1/2} \tag{2.53}$$

where A is an absolute constant. The lemma now follows from (2.52) and (2.53). □

We give now some corollaries of Theorem 2.6.

COROLLARY 2.7. *Let $f \in BV(\mathbf{T})$ with $\int_{\mathbf{T}} f(t)dt = 0$. Then $\sum_{k=1}^{\infty} c_k f(kx)$ converges a.e. provided*

$$\sum_{k=1}^{\infty} c_k^2 (\log k)^{\beta} < \infty \quad \textit{for some } \beta > 2. \tag{2.54}$$

COROLLARY 2.8. *Let $f \in \mathrm{Lip}_{\alpha}(\mathbf{T})$ for some $0 < \alpha < 1/2$ and let $\int_{\mathbf{T}} f(t)dt = 0$. Then $\sum_{k=1}^{\infty} c_k f(kx)$ converges a.e. provided*

$$\sum_{k=1}^{\infty} c_k^2 k^{1-2\alpha} (\log k)^{\beta} < \infty \qquad \textit{for some } \beta > 1 + 2\alpha. \tag{2.55}$$

COROLLARY 2.9. *Let $f \in \mathrm{Lip}_{1/2}(\mathbf{T})$ and let $\int_{\mathbf{T}} f(t)dt = 0$. Then $\sum_{k=1}^{\infty} c_k f(kx)$ converges a.e. provided*

$$\sum_{k=1}^{\infty} c_k^2 (\log k)^{\beta} < \infty \quad \textit{for some } \beta > 2. \tag{2.56}$$

Corollary 2.8 was proved earlier by Gaposhkin ([14], Theorem 4), while Corollary 2.9 improves Theorem 3 of Gaposhkin [14] in the $\mathrm{Lip}_{1/2}(\mathbf{T})$ case.

Note that the convergence condition (2.55), valid under $0 < \alpha < 1/2$, is much more restrictive than condition (2.56), valid under $\alpha = 1/2$. It is possible that in the case $0 < \alpha < 1/2$ the condition

$$\sum_{k=1}^{\infty} c_k^2 (\log k)^{\gamma} < \infty \tag{2.57}$$

for a suitable $\gamma > 0$ suffices for the a.e. convergence of $\sum_{k=1}^{\infty} c_k f(kx)$, but this remains open. On the other hand, Theorem 3 of Berkes [4] shows that for any $0 < \alpha < 1/2$ there exists a function $f \in \mathrm{Lip}_{\alpha}(\mathbf{T})$ with $\int_{\mathbf{T}} f(t)dt = 0$ and a real sequence (c_k) such that (2.57) holds for all $\gamma < 1 - 2\alpha$, but $\sum_{k=1}^{\infty} c_k f(kx)$ is a.e. divergent.

To prove the corollaries, assume first that $f \in \mathrm{Lip}_{\alpha}(\mathbf{T})$ with some $0 < \alpha \leq 1/2$. (As we noted above, in the case $\alpha > 1/2$ the series $\sum_{k=1}^{\infty} c_k f(kx)$ converges a.e. for any $\mathbf{c} \in \ell^2$ by Gaposhkin's theorem, so there is no convergence problem.) The Fourier coefficients of f satisfy (see Zygmund [55], Vol I, p. 241)

$$\sum_{k=2^n+1}^{2^{n+1}} (a_k^2 + b_k^2) \leq C 2^{-2n\alpha}$$

whence it follows immediately that

$$\sum_{k=n}^{\infty} (a_k^2 + b_k^2) \leq C n^{-2\alpha} \tag{2.58}$$

and

$$\sum_{k=1}^{\infty} (|a_k| + |b_k|) k^{\alpha - 1/2} (\log k)^{-\gamma} < \infty \qquad \text{for any } \gamma > 1. \tag{2.59}$$

The cases $0 < \alpha < 1/2$ and $\alpha = 1/2$ are treated differently, so we separate them.

(A) In the case $\alpha = 1/2$ we note that $r_f(n) = O(n^{-1})$ by (2.58) and thus by (2.56) and the Cauchy–Schwarz inequality the first term on the right side of (2.50) is bounded by

$$C \sum_{k \leq N} |c_k| \frac{1}{\sqrt{m_k}} = C \sum_{k \leq N} |c_k| (\log k)^{\beta/2} \frac{1}{\sqrt{m_k} (\log k)^{\beta/2}} \leq C \left(\sum_{k \leq N} \frac{1}{m_k (\log k)^{\beta}} \right)^{1/2}$$

which remains bounded if $m_k = k(\log k)^{1+\varepsilon-\beta}$, $\varepsilon > 0$. Then $d_l \sim l(\log l)^{-(1+\varepsilon-\beta)}$ and since by (2.56) we have $\sum_{k \geq N} c_k^2 \leq C(\log N)^{-\beta}$, the second term on the right-hand side of (2.50) is bounded by

$$C \sum_{l=1}^{m_N} (|a_l| + |b_l|)(\log d_l)^{-\beta/2}$$

which remains bounded by (2.59), since $\log d_l \sim \log l$ and $\beta > 2$. Thus Corollary 2.9 is proved.

Observe that if f is of bounded variation, then its Fourier coefficients satisfy $|a_k| = O(k^{-1})$, $|b_k| = O(k^{-1})$, and thus relations (2.58), (2.59) are satisfied with $\alpha = 1/2$. Hence the above proof also shows the validity of Corollary 2.7.

(B) In the case of Corollary 2.8, i.e. $0 < \alpha < 1/2$ we choose $m_k = k(\log k)^\tau$ with τ to be determined later; then $d_l \sim l(\log l)^{-\tau}$. By (2.58) we have $r_f(n) = O(n^{-2\alpha})$ and thus setting $\psi(k) = k^{1-2\alpha}(\log k)^\beta$, (2.55) and the Cauchy–Schwarz inequality show that the first term on the right side of (2.50) is bounded by

$$C\sum_{k\le N}|c_k|\frac{1}{m_k^\alpha} = C\sum_{k\le N}|c_k|\psi(k)^{1/2}\frac{1}{m_k^\alpha\psi(k)^{1/2}} \le C\left(\sum_{k\le N}\frac{1}{m_k^{2\alpha}\psi(k)}\right)^{1/2}$$

which remains bounded, in view of the definitions of m_k and $\psi(k)$, if $\beta + 2\alpha\tau > 1$. On the other hand, $\sum_{k=1}^\infty c_k^2\psi(k) < \infty$ implies $\sum_{k\ge N} c_k^2 \le C\psi(N)^{-1}$, and thus the second term on the right-hand side of (2.50) is bounded by

$$C\sum_{l=1}^{m_N}(|a_l| + |b_l|)\psi(d_l)^{-1/2}. \tag{2.60}$$

Substituting the values of $\psi(k)$ and d_l and using (2.59), we see that the sum in (2.60) remains bounded if $\beta - (1-2\alpha)\tau > 2$. We thus proved that if the sum in (2.55) converges and $m_k = k(\log k)^\tau$, then the right-hand side of (2.50) remains bounded if

$$\beta > \max\big(2 + (1-2\alpha)\tau, 1 - 2\alpha\tau\big). \tag{2.61}$$

The right-hand side (2.61) reaches its minimum for $\tau = -1$ with minimal value $1 + 2\alpha$, completing the proof. □

CHAPTER 3

Almost everywhere convergence: necessary conditions

Let $f \in L^2(\mathbf{T})$ with $\int_{\mathbf{T}} f(t)dt = 0$ and Fourier expansion

$$f \sim \sum_{k=1}^{\infty} (a_k \cos 2\pi kx + b_k \sin 2\pi kx).$$

Recall that by Wintner's theorem (Theorem A in Ch. 1), the series $\sum_{n=1}^{\infty} c_n f(nx)$ converges in the mean for all $\mathbf{c} \in \ell^2$ iff

$$\sum_{n=1}^{\infty} a_n/n^s \quad \text{and} \quad \sum_{n=1}^{\infty} b_n/n^s \quad \text{are regular and bounded for } \Re s > 0. \tag{3.1}$$

In Chapter 2 we proved (see Corollary 2.2) that under minor smoothness assumptions, (3.1) also implies the a.e. convergence of $\sum_{k=1}^{\infty} c_k f(n_k x)$ provided (n_k) satisfies the Erdős gap condition (2.11) with $\beta < 1/2$. The following result describes the situation when (3.1) fails.

THEOREM 3.1. *Let $f \in \mathrm{Lip}_\alpha(\mathbf{T})$, $0 < \alpha \leq 1$, $\int_{\mathbf{T}} f(t)dt = 0$ and assume that (3.1) is not valid. Then for any $\varepsilon_k \downarrow 0$ there exists $\mathbf{c} \in \ell^2$ and a sequence (n_k) of positive integers satisfying*

$$n_{k+1}/n_k \geq 1 + \varepsilon_k \qquad (k \geq k_0)$$

such that the series $\sum_k c_k f(n_k x)$ is a.e. divergent.

This result is sharp: if (n_k) grows exponentially (i.e. $n_{k+1}/n_k \geq q > 1$), then $\sum_{k=1}^{\infty} c_k f(n_k x)$ converges a.e. for any $\mathbf{c} \in \ell^2$ by Kac's theorem (see Theorem D in Chapter 2).

We note that Theorem 3.1 remains valid, with minor modifications in the proof, if instead of $f \in \mathrm{Lip}_\alpha(\mathbf{T})$ we assume only $f \in L^2(\mathbf{T})$. However, since the positive result was obtained under smoothness conditions, we will prove the converse also for this case.

For the proof we need two simple lemmas.

LEMMA 3.1. *If (3.1) fails, then for any $N \geq 1$ there exist real numbers $\{a_j^{(N)}, j = 1, \dots, N\}$ such that*

$$\int_0^1 \left(\sum_{j=1}^{N} a_j^{(N)} f(jx) \right)^2 dx \geq \left(\sum_{j=1}^{N} (a_j^{(N)})^2 \right) L(N)$$

where $L(N) \to \infty$.

PROOF. This is obvious, since by Wintner's theorem relation (3.1) is equivalent to the existence of a constant $C > 0$ such that for any $N \geq 1$ and any real sequence $(a_j)_{1 \leq j \leq N}$ we have

$$\int_0^1 \Big(\sum_{j=1}^N a_j f(jx) \Big)^2 dx \leq C \Big(\sum_{j=1}^N a_j^2 \Big).$$

□

Now, given $f \in \mathrm{Lip}_\alpha(\mathbf{T})$, choose the integer B so large that $B\alpha \geq 4$. Then we have

LEMMA 3.2. *Let $1 \leq p_1 < q_1 < p_2 < q_2 < \dots$ be integers such that $p_{k+1} \geq Bq_k$. Let $I_1, I_2, \dots$ be sets of integers such that $I_k \subset [2^{p_k}, 2^{q_k}]$ and each element of I_k is divisible by 2^{p_k}. Let $b_j^{(k)}$, $j \in I_k$ be arbitrary coefficients with $|b_j^{(k)}| \leq 1$ and set*

$$X_k = X_k(\omega) = \sum_{j \in I_k} b_j^{(k)} f(j\omega) \qquad (k = 1, 2, \dots, \ \ \omega \in [0, 1)).$$

Then there exist independent r.v.'s $Y_1, Y_2, \dots$ on the probability space $([0, 1), \mathcal{B}, \lambda)$ such that $\mathbf{E}\, Y_k = 0$ and

$$|X_k - Y_k| \leq 2^{-k} \qquad (k \geq k_0).$$

PROOF. Let $\mathcal{F}_k$ denote the σ-field generated by the dyadic intervals

$$U_\nu = \big[\nu 2^{-Bq_k}, (\nu+1) 2^{-Bq_k}\big] \qquad 0 \leq \nu \leq 2^{Bq_k} - 1 \tag{3.2}$$

and set

$$\xi_j = \xi_j(\cdot) = \mathbf{E}\,(f(j \cdot) | \mathcal{F}_k), \qquad j \in I_k$$

$$Y_k = Y_k(\omega) = \sum_{j \in I_k} b_j^{(k)} \xi_j(\omega).$$

By $|f(x) - f(y)| \leq C|x - y|^\alpha$ we have, using $B\alpha \geq 4$,

$$|\xi_j(\omega) - f(j\omega)| \leq C_1 \cdot 2^{q_k} 2^{-Bq_k \alpha} \leq C_1 \cdot 2^{-3q_k} \qquad j \in I_k$$

and since I_k has at most 2^{q_k} elements, we get

$$|X_k - Y_k| \leq C_1 \cdot 2^{-2q_k} \leq 2^{-k} \quad \text{for} \quad k \geq k_0.$$

Since $p_{k+1} \geq Bq_k$ and since each $j \in I_{k+1}$ is a multiple of $2^{p_{k+1}}$, each interval U_ν in (3.2) is a period interval for all $f(jx)$, $j \in I_{k+1}$ and thus also for ξ_j, $j \in I_{k+1}$. Hence Y_{k+1} is independent of the σ-field $\mathcal{F}_k$ and since $\mathcal{F}_1 \subset \mathcal{F}_2 \subset \dots$ and Y_k is $\mathcal{F}_k$ measurable, the r.v.'s $Y_1, Y_2, \dots$ are independent. Finally $E\xi_j = 0$ by $\int_{\mathbf{T}} f dx = 0$ and thus $EY_k = 0$.

Turning to the proof of Theorem 3.1, let $\psi(k)$ grow so rapidly that $L(\psi(k)) \geq 2^k$ and let (r_k) be a nondecreasing sequence of integers to be chosen later. We define sets

$$I_1^{(1)}, I_2^{(1)}, \dots, I_{r_1}^{(1)}, I_1^{(2)}, \dots, I_{r_2}^{(2)}, \dots, I_1^{(k)}, \dots, I_{r_k}^{(k)}, \dots \tag{3.3}$$

of positive integers by

$$I_j^{(k)} = 2^{c_j^{(k)}} \{1, 2, \dots, \psi(k)\}, \qquad 1 \leq j \leq r_k, \ k \geq 1$$

where $c_j^{(k)}$ are suitable positive integers. (Here for any set $\{a, b, \dots\} \subset \mathbf{R}$ and $\lambda \in \mathbf{R}$, $\lambda\{a, b, \dots\}$ denotes the set $\{\lambda a, \lambda b, \dots\}$.) Clearly we can choose the integers $c_j^{(k)}$ inductively so that the intervals in (3.3) satisfy the conditions of Lemma 3.2. By Lemma 3.1 there exist, for any $k \ge 1$, coefficients $\{a_\nu^{(k)},\ 1 \le \nu \le \psi(k)\}$, $\sum_{\nu=1}^{\psi(k)} a_\nu^{(k)2} = 1$ such that, setting

$$X^{(k)} = X^{(k)}(\omega) = \sum_{\nu=1}^{\psi(k)} a_\nu^{(k)} f(\nu\omega)$$

we have

$$\mathbf{E}\left(X^{(k)}\right)^2 \ge L(\psi(k)).$$

Let

$$X_j^{(k)}(\omega) = X^{(k)}(2^{c_j^{(k)}}\omega), \qquad 1 \le j \le r_k.$$

Clearly the $X_j^{(k)}$ have the same distribution, and consequently

$$\mathbf{E}\left(X_j^{(k)}\right)^2 \ge L(\psi(k)).$$

By Lemma 3.1 there exist independent r.v.'s $Y_j^{(k)}$ $(1 \le j \le r_k,\ k = 1, 2, \dots)$ such that $\mathbf{E}\,Y_j^{(k)} = 0$ and

$$\sum_{k,j} |X_j^{(k)} - Y_j^{(k)}| \le K \tag{3.4}$$

for some constant $K > 0$. Hence by the Minkowski inequality

$$\mathbf{E}(Y_j^{(k)})^2 \ge \frac{1}{2} L(\psi(k)) \tag{3.5}$$

for $k \ge k_0$. Also $|Y_j^{(k)}| \le |X_j^{(k)}| + K \le \text{Const} \cdot \psi(k)$ and thus setting

$$Z_k = \frac{1}{(r_k L\psi(k))^{1/2}} \sum_{j=1}^{r_k} Y_j^{(k)}$$

$$\sigma_k^2 = \mathbf{E}\left(\sum_{j=1}^{r_k} Y_j^{(k)}\right)^2 \ge \frac{1}{2} r_k L(\psi(k))$$

we get from the central limit theorem with Berry–Esseen remainder term

$$\begin{aligned}\mathbf{P}\{Z_k \ge 1\} \ge \mathbf{P}\left\{\sum_{j=1}^{r_k} Y_j^{(k)} \ge 2\sigma_k\right\} &\ge (1 - \Phi(2)) - C r_k \psi(k)^3 (r_k L(\psi(k))^{-3/2} \\ &\ge 1 - \Phi(2) - o(1) \ge 0.02 \qquad (k \ge k_0)\end{aligned}$$

provided r_k grows so rapidly that $r_k^{1/2} L(\psi(k))^{3/2} \ge \psi(k)^4$. Since the Z_k are independent, the Borel–Cantelli lemma implies $\mathbf{P}\{Z_k \ge 1 \text{ i.o.}\} = 1$, i.e. $\sum_{k \ge 1} Z_k$ is a.e. divergent, which, in view of (3.4), yields that

$$\sum_{k=1}^{\infty} \frac{1}{(r_k L(\psi(k))^{1/2}} \sum_{j=1}^{r_k} X_j^{(k)} \qquad \text{is a.e. divergent.} \tag{3.6}$$

Let now

$$(n_k) = \bigcup_{k=1}^{\infty} \bigcup_{j=1}^{r_k} I_j^{(k)}. \tag{3.7}$$

Then the sum in (3.6) is of the form $\sum_{i=1}^{\infty} c_i f(n_i x)$ where

$$\sum_{i=1}^{\infty} c_i^2 = \sum_{k=1}^{\infty} \frac{r_k}{r_k L(\psi(k))} = \sum_{k=1}^{\infty} \frac{1}{L(\psi(k))} < +\infty.$$

Finally, denote by $1 + \rho_k$ the smallest of the ratios $(j+1)/j$, $1 \le j \le \psi(k) - 1$; clearly $\rho_k > 0$. Given $\varepsilon_k \downarrow 0$ one can choose r_k growing so rapidly that

$$\rho_k \ge \varepsilon_{r_{k-1}} \qquad k = 1, 2, \ldots. \tag{3.8}$$

Now if n_s and n_{s+1} belong to the same set $I_j^{(k)}$ then clearly $s \ge r_{k-1}$ and thus by (3.8) we get $n_{s+1}/n_s \ge 1 + \rho_k \ge 1 + \varepsilon_{r_{k-1}} \ge 1 + \varepsilon_s$. Since $n_{s+1}/n_s \ge 2$ if n_s and n_{s+1} belong to different $I_j^{(k)}$'s, we proved that (n_k) satisfies

$$n_{k+1}/n_k \ge 1 + \varepsilon_k \qquad (k \ge k_0). \tag{3.9}$$

This completes the proof of Theorem 3.1. □

There are few results concerning the bounded case, namely the case when in the series $\sum_{k=1}^{\infty} c_k f(n_k x)$, f is not smooth but only bounded. We first consider the case of primes and prove the following result.

THEOREM 3.2. *Let $\mathcal{P} := (P_k)$ be an increasing sequence of prime numbers. Let $\mathbf{c} = \{c_k, k \ge 1\}$ be a sequence of positive reals such that*

$$\sum_k c_k^2 < \infty, \qquad \sum_k c_k = \infty.$$

Then there exists a function $f \in L^\infty(\mathbf{T})$ with $\int_{\mathbf{T}} f(t)dt = 0$ such that the series $\sum_{k=1}^{\infty} c_k f(P_k x)$ diverges on a set with positive measure.

Theorem 3.2 will be deduced from the following

THEOREM 3.3. *Let $\mathcal{P} := (P_k)$ be an increasing sequence of prime numbers. Let $\mathbf{c} = \{c_k, k \ge 1\}$ be a sequence of positive reals such that*

$$\sum_k c_k^2 < \infty, \qquad \sum_k c_k = \infty.$$

Put $C_n = \sum_{k \le n} c_k$ and consider the weighted sums

$$\mathcal{S}_n f = \frac{1}{C_n} \sum_{k \le n} c_k f(P_k x).$$

Then there exists a function $f \in L^\infty(\mathbf{T})$ with $\int_{\mathbf{T}} f(t)dt = 0$ such that the sequence $\{\mathcal{S}_n f, n \ge 1\}$ diverges on a set with positive measure.

PROOF OF THEOREM 3.2. Assuming that Theorem 3.3 is valid, there exists a bounded measurable function f such that $(\mathcal{S}_n f)_n$ does not converge almost everywhere. Then the partial sums $\sum_{k\leq n} c_k f(P_k x)$ do not converge almost everywhere either. Otherwise, this would imply, in view of the assumption that the series $\sum_k c_k$ diverges, that $(\mathcal{S}_n f(x))_n$ tend to 0 almost everywhere, a contradiction. Hence the result. □

To prove Theorem 3.3, we use Bourgain's entropy criterion in L^∞ which we recall here.

LEMMA 3.3 (Bourgain [7], Proposition 2]). *Let $\{S_n, n \geq 1\}$ be a sequence of $L^2(\mu) - L^\infty(\mu)$ contractions satisfying the following commutation assumption:*

(H) *There exists a family $\mathcal{E} = \{T_j, j \geq 1\}$ of μ-preserving measurable transformations of X, commuting with S_n $((S_n T_j(f) = T_j S_n(f))$ such that for any $g \in L^1(\mathbf{T})$,*

$$\lim_{J\to\infty} \left\| \frac{1}{J}\sum_{j=1}^{J} T_j g - \int_{\mathbf{T}} g d\lambda \right\|_1 = 0. \tag{3.10}$$

Moreover, assume that

$$\mu\{S_n(f) \text{ converges as } n \to \infty\} = 1 \quad \text{for all } f \in L^\infty(\mu). \tag{3.11}$$

For any $\delta > 0$, let $N_f(\delta)$ denotes the minimal number of $L^2(\mu)$-open balls centered in the set $\{S_n f, n \geq 1\}$ and enough to cover it. Then,

$$C(\delta) = \sup_{f\in L^\infty(\mu),\ ||f||_2=1} N_f(\delta) < \infty. \tag{3.12}$$

If the T_j's are defined as at the beginning of Section 2, we know that $S_n T_j(f) = T_j S_n(f)$ and for any $g \in L^2(\mathbf{T})$, $\lim_{J\to\infty} \left\| \frac{1}{J}\sum_{j=1}^{J} T_j g - \int_{\mathbf{T}} g d\lambda \right\|_2 = 0$. This, plus a plain approximation argument and the fact that barycenters of contractions are again contractions, finally imply (3.10). This means that assumption **(H)** is satisfied in our case. For proving Theorem 3.3, we will also need the lemma below, which is taken from Weber [48], (see Lemma 5.1.6 on p. 76).

LEMMA 3.4. *Let R, T, p be three positive integers such that $R \geq T(4p^2 - 3)$. Let $(H, ||.||)$ be a Hilbert space. Let $B = \{f_n,\ 1 \leq n \leq R\}$ be a finite subset of H such that $||f_n|| \leq 1$, $1 \leq n \leq R$ and $\Phi = \{\phi_n,\ 1 \leq n \leq R\}$ an orthonormal system of H. We assume*

$$\langle f_n, \phi_n \rangle \geq \frac{1}{p} \qquad (1 \leq n \leq R) \tag{a}$$

Then B contains a subset B' satisfying

$$\text{Card}\,(B') \geq T \quad \textit{and} \quad \inf_{f,g\in B',\ f\neq g} ||f - g|| \geq \frac{1}{2p}. \tag{b}$$

PROOF OF THEOREM 3.3. Let $\{T_N, N \geq 1\}$ be integers such that $T_N - T_{N-1}$ increases to infinity with N. Define

$$\Pi_N = \{u = P_{T_{N-1}+1}^{\alpha_{T_{N-1}+1}} \dots P_{T_N}^{\alpha_{T_N}} : \alpha_i \in \{0,1\} \text{ and } (\alpha_{T_{N-1}+1}, \dots, \alpha_{T_N}) \neq (0, \dots, 0)\},$$

$$f_N = \frac{1}{\left[2^{T_N - T_{N-1}} - 1\right]^{1/2}} \sum_{u \in \Pi_N} e_u. \tag{3.13}$$

Let $T_{N-1} < R \leq T_N$. Then,

$$\langle \mathcal{S}_R(f_N), f_N \rangle = \frac{1}{C_R} \frac{1}{\left[2^{T_N - T_{N-1}} - 1\right]^{1/2}} \sum_{u \in \Pi_N} \sum_{v \in \Pi_N} \sum_{k \leq R} c_k \langle e_{uP_k}, e_v \rangle.$$

Let $u, v \in \Pi_N$ and $k \leq R$. Then $\langle e_{uP_k}, e_v \rangle = 1$, if and only if $uP_k = v$. Noting $u = P_{T_{N-1}+1}^{\alpha_{T_{N-1}+1}} \dots P_{T_N}^{\alpha_{T_N}}$, $v = P_{T_{N-1}+1}^{\beta_{T_{N-1}+1}} \dots P_{T_N}^{\beta_{T_N}}$, this means that:

$$P_k P_{T_{N-1}+1}^{\alpha_{T_{N-1}+1}} \dots P_{T_N}^{\alpha_{T_N}} = P_{T_{N-1}+1}^{\beta_{T_{N-1}+1}} \dots P_{T_N}^{\beta_{T_N}}.$$

This equation has solutions if and only if k belongs to the interval $]T_{N-1}, T_N]$, and then the solutions are given by

$$\alpha_k = 0, \quad \beta_k = 1 \qquad \alpha_j = \beta_j, \qquad \textit{otherwise}.$$

Hence,

$$\langle f_N(P_k \cdot), f_N \rangle = \frac{2^{T_\theta - T_{\theta-1} - 1} - 1}{2^{T_\theta - T_{\theta-1}} - 1} \geq \frac{1}{4}. \tag{3.14}$$

Consequently, for any integer $N \geq 1$ and any $T_{N-1} < R \leq T_N$

$$\begin{aligned} &\langle \mathcal{S}_R(f_N), f_N \rangle \\ &= \frac{1}{C_R} \sum_{k \leq R} c_k \langle f_N(P_k \cdot), f_N \rangle = \frac{1}{C_R} \sum_{\substack{k \leq R \\ k \in]T_{N-1}, T_N]}} c_k \langle f_N(P_k \cdot), f_N \rangle \geq \frac{1}{4}. \end{aligned} \tag{3.15}$$

The proof is completed by applying Lemma 3.4 and the entropy criterion in L^∞. □

The next two theorems will concern subsequences $\mathcal{N}$ generated by infinitely many primes.

THEOREM 3.4. *Let $\mathcal{P} = \{P_1, P_2, \cdots\}$ be an increasing sequence of positive pairwise coprime integers, and denote by $\mathcal{C}(\mathcal{P})$ the infinite dimensional chain generated by $\mathcal{P}$. Let $\mathbf{c} = \{c_k, k \geq 1\}$ be a sequence of positive reals such that the series $\sum_{k=1}^\infty c_k$ diverges. Define for any measurable function $f : \mathbf{T} \to \mathbf{R}$ the weighted sums*

$$\mathcal{S}_n f(x) = \frac{1}{\sum_{j \in \mathcal{C}(\mathcal{P}) \cap [1,n]} c_j} \sum_{j \in \mathcal{C}(\mathcal{P}) \cap [1,n]} c_j f(jx).$$

Assume that

$$\limsup_{i \to \infty} \frac{\sum_{j \in \mathcal{C}(\mathcal{P}) \cap [\frac{1}{2} P_1^{2i}, P_1^{2i}[)} c_j}{\sum_{j \in \mathcal{C}(\mathcal{P}) \cap [1, P_1^{2i}[)} c_j} > 0. \tag{3.16}$$

Then there exists a bounded measurable function f such that $(\mathcal{S}_n f)_n$ does not converge almost everywhere.

From Theorem 3.4 one can obtain

THEOREM 3.5. *Let $\mathcal{P} = \{P_1, P_2, \cdots\}$ be an increasing sequence of positive pairwise coprime integers, and denote by $\mathcal{C}(\mathcal{P})$ the infinite dimensional chain generated by $\mathcal{P}$. Let $\mathbf{c} = \{c_k, k \geq 1\}$ be a sequence of positive reals such that*

$$\sum_k c_k^2 < \infty \qquad \sum_k c_k = \infty.$$

Assume that condition (3.16) is satisfied. Then, there exists a bounded measurable function f such that $\sum_{k=1}^{\infty} c_k f(P_k x)$ does not converge almost everywhere.

The proof of Theorem 3.5 is similar to the proof of Theorem 3.3, so it is omitted.

PROOF OF THEOREM 3.4. The proof uses Bourgain's ideas in [7]. Let s be some fixed positive integer. Put for any integer $T \geq 0$

$$A_T = \left\{n = P_1^{\alpha_1} \cdots P_s^{\alpha_s} \ : \ P_1^T \leq n < P_1^{T+1}, \ \alpha_i \geq 0, \ i = 1, \cdots, s\right\}. \tag{3.17}$$

By replacing α_1 by $\alpha_1 + 1$, one can easily verify that

$$\sharp(A_T) \leq \sharp(A_{T+1}) \tag{3.18}$$

As for $n = P_1^{\alpha_1} \cdots P_s^{\alpha_s} \in A_T$, necessarily $0 \leq \alpha_1 + \cdots + \alpha_s \leq T$, we also deduce

$$\sharp(A_T) \leq T^s. \tag{3.19}$$

Then, for any $d > 0$, there exists an integer $T > 0$ such that

$$\sharp(A_{T+d}) \leq 2\sharp(A_T), \tag{3.20}$$

Indeed, otherwise, $\sharp(A_{T+d}) > 2\sharp(A_T)$ for any T, would imply for any integer n

$$\sharp(A_{nd}) > B2^n,$$

where B is some positive constant, which contradicts to (3.19). Choose d such that $P_1^d \leq P_s$. Any element $j \in \mathcal{C}(\mathcal{P})$ such that $j \leq P_1^d$ can be thus expressed as $j = P_1^{\alpha_1} \cdots P_r^{\alpha_r}$ with $r \leq s$. Put for any $i = 0, \cdots, d$

$$f^{(i)}(x) = \frac{1}{\sharp(A_{T+i})^{\frac{1}{2}}} \sum_{n \in A_{T+i}} e^{2i\pi nx}, \tag{3.21}$$

and let

$$f = f^{(0)}.$$

Next, put for any $i = 0, \cdots, [\frac{d}{2}]$

$$\phi_i = \frac{f^{(2i-1)} + f^{(2i)}}{\sqrt{2}}, \tag{3.22}$$

and let for any integer j, $f_j(x) = f(jx)$. The set of functions $f^{(i)}$ is a sub-orthonormal system of L^2, the same property holds true for the system of functions ϕ_i. Moreover $||f_j|| = 1$ for any j. Let $1 \leq i \leq [\frac{d}{2}]$, $j \in [P_1^{2i-1}, P_1^{2i}] \cap \mathcal{C}(\mathcal{P})$, and examine f_j. Let $n \in A_T$. Then nj may be written as $nj = P_1^{\beta_1} \cdots P_s^{\beta_s}$. Moreover

$$P_1^{T+2i-1} \leq nj < P_1^{T+2i+1}.$$

It follows that we have the implication

$$n \in A_T \textit{ and } j \in [P_1^{2i-1}, P_1^{2i}] \cap \mathcal{C}(\mathcal{P}) \quad \Rightarrow \quad nj \in A_{T+2i-1} \cup A_{T+2i}.$$

We may thus write

$$f_j(x) = \frac{1}{\sharp(D)^{\frac{1}{2}}} \sum_{m\in D} e^{2i\pi mx},$$

where $D \subset A_{T+2i-1} \cup A_{T+2i}$ and $\sharp(D) = \sharp(A_T)$. Hence,

$$\begin{aligned}\sqrt{2}\langle f_j, \phi_i\rangle &= \frac{1}{[\sharp(A_T)\sharp(A_{T+2i-1})]^{\frac{1}{2}}} \sum_{m\in D\cap A_{T+2i-1}} 1 + \frac{1}{[\sharp(A_T)\sharp(A_{T+2i})]^{\frac{1}{2}}} \sum_{m\in D\cap A_{T+2i}} 1 \\ &\geq \frac{1}{\sharp(A_T)\sqrt{2}}.\sharp(A_T) = \frac{1}{\sqrt{2}},\end{aligned}$$

and so for any $1 \leq i \leq [\frac{d}{2}]$, $P_1^{2i-1} \leq j \leq P_1^{2i}$

$$\langle f_j, \phi_i\rangle \geq \frac{1}{2}. \tag{3.23}$$

Further, $\langle f_j, \phi_k\rangle \geq 0$ for any j and k. Thus,

$$\begin{aligned}\langle S_{P_1^{2i}}(f), \phi_i\rangle &= \frac{1}{\sum_{j\in\mathcal{C}(\mathcal{P})\cap[1,P_1^{2i}]} c_j} \sum_{j\in\mathcal{C}(\mathcal{P})\cap[1,P_1^{2i}])} c_j\langle f_j, \phi_i\rangle \\ &\geq \frac{1}{\sum_{j\in\mathcal{C}(\mathcal{P})\cap[1,P_1^{2i}]} c_j} \sum_{j\in\mathcal{C}(\mathcal{P})\cap[\frac{1}{2}P_1^{2i},P_1^{2i}])} c_j\langle f_j, \phi_i\rangle \\ &\geq \frac{1}{2}\frac{\sum_{j\in\mathcal{C}(\mathcal{P})\cap[\frac{1}{2}P_1^{2i},P_1^{2i}]} c_j}{\sum_{j\in\mathcal{C}(\mathcal{P})\cap[1,P_1^{2i}]} c_j}\end{aligned}$$

We have obtained for any $i = 1, \cdots, [\frac{d}{2}]$

$$\langle S_{P_1^{2i}}(f), \phi_i\rangle \geq \frac{1}{2}\frac{\sum_{j\in\mathcal{C}(\mathcal{P})\cap[\frac{1}{2}P_1^{2i},P_1^{2i}]} c_j}{\sum_{j\in\mathcal{C}(\mathcal{P})\cap[1,P_1^{2i}]} c_j}. \tag{3.24}$$

Now, by assumption

$$\limsup_{i\to\infty} \frac{\sum_{j\in\mathcal{C}(\mathcal{P})\cap[\frac{1}{2}P_1^{2i},P_1^{2i}]} c_j}{\sum_{j\in\mathcal{C}(\mathcal{P})\cap[1,P_1^{2i}]} c_j} > 0.$$

We may find an increasing sequence $(i_\lambda)_\lambda$ of integers as well as a positive real c, such that

$$\frac{\sum_{j\in\mathcal{C}(\mathcal{P})\cap[\frac{1}{2}P_1^{2i_\lambda},P_1^{2i_\lambda}]} c_j}{\sum_{j\in\mathcal{C}(\mathcal{P})\cap[1,P_1^{2i_\lambda}]} c_j} \geq 2c \qquad (\lambda = 1, 2\ldots)$$

Consequently, for any λ such that $i_\lambda \leq d$,

$$\langle S_{P_1^{2i_\lambda}}(f), \phi_{i_\lambda}\rangle \geq c. \tag{3.25}$$

Let p be a positive integer such that $pc \geq 1$. Lemma 3.4 applied with the choices $R = [\frac{D}{2}]$, $T = [[\frac{D}{2}]/13]$ with $D = \sharp(\lambda \mid i_\lambda \leq d)$ and p shows that

$$N\Big(\Big(S_{P_1^{2i}}(f), i \leq \Big[\frac{D}{2}\Big]\Big), \frac{c}{2}\Big) \geq T. \tag{3.26}$$

But d is arbitrary, thus

$$\sup_{f\in L^\infty \ ||f||_2\leq 1} N\left((S_{P_1^{2i}}(f), i \geq 1), \frac{c}{2}\right) = \infty.$$

Applying now Bourgain's entropy criterion in L^∞ concludes the proof. □

CHAPTER 4

Random sequences

In this chapter we investigate the convergence of the series $\sum_{k=1}^{\infty} c_k f(n_k x)$ where (n_k) is a random sequence of real numbers. Specifically, we will investigate the model when $n_k = X_1 + \cdots + X_k$, where the X_k are independent, identically distributed random variables defined on some probability space $(\Omega, \mathcal{A}, \mathbf{P})$. We will not assume that X_1 is integer valued or $X_1 > 0$; we assume only that the the distribution of X_1 is nondegenerate. If the random walk $\{\sum_{k=1}^{n} X_k,\ n \geq 1\}$ is transient, we have $|n_k| \to \infty$ a.s. On the other hand, if the random walk is recurrent and X_1 is nonlattice, (n_k) is dense in $\mathbf{R}$ with probability 1.

We begin our investigations with the study of random trigonometric sums of the form

$$\sum_{n=1}^{\infty} c_n e^{itS_n(\omega)} \tag{4.1}$$

where $\mathbf{c} \in \ell^2$; the terms of this sum are functions defined on the product space $\Omega \times \mathbf{T}$, endowed with the product probability $\mathbf{P} \times \lambda$.

THEOREM 4.1. *Let X_1 be nondegenerate with characteristic function φ and let $S_n = \sum_{k=1}^{n} X_k$ be the corresponding random walk. Then for any $\mathbf{c} \in \ell^2$ and any real t for which*

$$\rho = \max\big(|\varphi(t)|, |\varphi(2t)|, |\varphi(-t)|, |\varphi(-2t)|\big) < 1 \tag{4.2}$$

the series (4.1) converges with probability 1. Consequently, the series (4.1) converges for almost all $(t, \omega) \in \mathbf{T} \times \Omega$, provided $\mathbf{c} \in \ell^2$.

Since X_1 is nondegenerate, (4.2) holds for all but countably many t's. If X_1 is nonlattice, then $|\varphi(t)| < 1$ for all $t \neq 0$; otherwise there exists a $t_0 > 0$ such that $|\varphi(t)| = 1$ if and only if $t = kt_0$, $k \in \mathbf{Z}$. If X_1 is degenerate, then $S_n = cn$ with some constant c, and the statement of Theorem 4.1 reduces to Carleson's theorem, which is of course not contained in our result. But it is interesting to note that for all other random walks, the above formulated "random" version of Carleson's theorem is valid. This seems paradoxical at first sight, since the random walk S_n can be recurrent, e.g. it is possible that $S_n = 0$ for infinitely many n. However, by the theory of random walks the set $H = \{n : S_n = 0\}$ is thin (it has roughly $O(n^{1/2})$ elements in the interval $[0, n]$) and Theorem 4.1 shows that $\sum_{k \in H} |c_k| < \infty$ even if $\sum_{k=1}^{\infty} |c_k| = \infty$.

For the proof of Theorem 4.1 we will need the following convergence result of probability theory, first observed by Stechkin in the context of orthogonal series (see e.g. Gaposhkin [13, pp. 29–31]). For the present version, see Billingsley [5], p. 102; for an alternative proof see Weber [49], Theorem 2.1.

LEMMA 4.1. *Let $\{\xi_i, i \geq 1\}$ be a sequence of random variables satisfying the assumption*

$$\mathbf{E}\Big|\sum_{i \leq l \leq j} \xi_l\Big|^{\gamma} \leq \Big(\sum_{i \leq l \leq j} u_l\Big)^{\alpha}, \qquad 0 \leq i \leq j < \infty,$$

where $\{u_i, i \geq 1\}$ is a sequence of nonnegative reals such that the series $\sum_{l=1}^{\infty} u_l$ converges and $\alpha > 1$, $\gamma > 0$. Then the series $\sum_{l=1}^{\infty} \xi_l$ converges almost surely. Moreover, for $\alpha > 1$, we have

$$\left\| \sup_{i,j\geq 1} \left| \sum_{i\leq l\leq j} \xi_l \right| \right\|_{\gamma} \leq C \Big(\sum_{l=1}^{\infty} u_l \Big)^{\alpha/\gamma},$$

where the constant C depends on α only.

Applying Lemma 4.1 with $\gamma = 4$, $\alpha = 2$, $u_k = c_k^2$, for proof of Theorem 4.1 it suffices to prove the following

LEMMA 4.2. *For any real $c_1, \dots, c_N$ and any real t we have*

$$\mathbf{E}\left|\sum_{k=1}^{N} c_k e^{itS_k}\right|^4 \leq \frac{1}{(1-\rho)^2}\Big(\sum_{k=1}^{N} c_k^2\Big)^2. \tag{4.3}$$

where ρ is defined by (4.2).

PROOF. In the case $\rho = 1$ the lemma is obvious, so we can assume $\rho < 1$. Clearly for any real $c_1, \dots, c_N$ we have

$$\mathbf{E}\left|\sum_{k=1}^{N} c_k e^{itS_k}\right|^4 = \sum_{1\leq j,k,l,m\leq N} c_j c_k c_l c_m \mathbf{E} e^{it(S_j - S_k + S_l - S_m)}. \tag{4.4}$$

We now claim that

$$\left|\mathbf{E} e^{it(\pm S_j \pm S_k \pm S_l \pm S_m)}\right| \leq \rho^{(|j-k|+|l-m|)} \qquad (j \geq k \geq l \geq m). \tag{4.5}$$

where in the last exponent there are two positive and two negative signs. Clearly we can assume that the sign of S_j in (4.5) is positive; otherwise we replace t by $-t$. There are 3 cases:

$$\begin{aligned}
\text{(a)}\quad \left|\mathbf{E} e^{it(S_j - S_k + S_l - S_m)}\right| &= \left|\mathbf{E} e^{it(S_j - S_k)}\right|\left|\mathbf{E} e^{it(S_l - S_m)}\right| = \left|\varphi(t)\right|^{j-k}\left|\varphi(t)\right|^{l-m} \\
&\leq \rho^{(|j-k|+|l-m|)}, \\
\text{(b)}\quad \left|\mathbf{E} e^{it(S_j - S_k - S_l + S_m)}\right| &= \left|\mathbf{E} e^{it(S_j - S_k)}\right|\left|\mathbf{E} e^{-it(S_l - S_m)}\right| = \left|\varphi(t)\right|^{j-k}\left|\varphi(-t)\right|^{l-m} \\
&\leq \rho^{(|j-k|+|l-m|)}, \\
\text{(c)}\quad \left|\mathbf{E} e^{it(S_j + S_k - S_l - S_m)}\right| &= \left|\mathbf{E} e^{it(S_j - S_k) + 2it(S_k - S_l) + it(S_l - S_m)}\right| \\
&= \left|\varphi(t)\right|^{j-k}\left|\varphi(2t)\right|^{k-l}\left|\varphi(t)\right|^{l-m} \leq \rho^{(|j-k|+|l-m|)},
\end{aligned}$$

proving (4.5). Thus splitting the sum on the right-hand side of (4.4) into 24 subsums corresponding to a fixed relative order of j, k, l, m and in each such sum renaming the indices j, k, l, m so that they will be nonincreasing in the renamed order, we get

$$\mathbf{E}\left|\sum_{k=1}^{N} c_k e^{itS_k}\right|^4 \leq 24 \sum_{N\geq j\geq k\geq l\geq m\geq 1} |c_j||c_k||c_l||c_m| \rho^{(|j-k|+|l-m|)}. \tag{4.6}$$

Summing the right-hand side of (4.6) first for those indices (j, k, l, m) for which $j - k = r$ and $l - m = s$ are fixed, we get by Cauchy's inequality

$$\begin{aligned}
&\sum_{1\le k,k+r,m,m+s\le N} |c_k||c_{k+r}||c_m||c_{m+s}|\rho^{r+s} \\
&\le \rho^{r+s} \sum_{1\le k,k+r\le N} |c_k||c_{k+r}| \sum_{1\le m,m+s\le N} |c_m||c_{m+s}| \\
&\le \rho^{r+s}\Big(\sum_{1\le k\le N} c_k^2\Big)^{1/2}\Big(\sum_{1\le k+r\le N} c_{k+r}^2\Big)^{1/2}\Big(\sum_{1\le m\le N} c_m^2\Big)^{1/2}\Big(\sum_{1\le m+s\le N} c_{m+s}^2\Big)^{1/2} \\
&\le \rho^{r+s}\Big(\sum_{1\le j\le N} c_j^2\Big)^2.
\end{aligned}$$

Now summing for r and s we get Lemma 4.2.

We turn now to the convergence of the series (4.1) in $L^p(\mathbf{T}\times\Omega)$ for $p > 2$. For simplicity, we consider the case $p = 4$. We will use the fact (Spitzer [46], p. 324) that for a transient integer valued random walk $\{S_n,\ n \ge 0\}$, the Green function

$$G(0,x) = \sum_{k=0}^{\infty} \mathbf{P}\{S_k = x\}$$

is finite for every $x \in \mathbf{Z}$. In particular

$$G(0,0) < \infty.$$

PROPOSITION 4.1. *Let $\mathcal{X} = \{X, X_i, i \ge 1\}$ be a sequence of independent, identically distributed, lattice random variables defined on some probability space $(\Omega, \mathcal{A}, \mathbf{P})$. We assume that the random walk $S_n = X_1 + \ldots + X_n$, $n \ge 1$ is transient. Then,*

$$\begin{aligned}
\mathbf{E}\int_{\mathbf{T}}\Big|\sum_{k=1}^{n} c_k\, e^{2\imath\pi\alpha S_k}\Big|^4 d\alpha &\le 4G(0,0)\Big(\sum_{k=1}^{n}|c_k|^2\Big) \\
&+ 6 \sum_{1\le i\le k<l\le j\le n} |c_i||c_j||c_k||c_l| \\
&\times \big\{\mathbf{P}\{S_k - S_i = \pm(S_j - S_l)\} + \mathbf{P}\{S_k - S_i = (S_j - S_l) - 2(S_l - S_k)\}\big\}
\end{aligned}$$

PROOF. Let $a_1, \dots, a_n$ be complex numbers. Then,

$$\begin{aligned}\Big|\sum_{i=1}^n a_i\Big|^4 &= \Big(\sum_{i=1}^n\sum_{j=1}^n a_i\bar a_j\Big)\Big(\sum_{k=1}^n\sum_{l=1}^n a_k\bar a_l\Big)\\ &= \Big(\sum_{i=1}^n |a_i|^2+\sum_{i=1}^n\sum_{j=i+1}^n (a_i\bar a_j+\bar a_i a_j)\Big)\Big(\sum_{k=1}^n |a_k|^2+\sum_{k=1}^n\sum_{l=k+1}^n (a_k\bar a_l+\bar a_k a_l)\Big)\\ &= \Big(\sum_{k=1}^n |a_k|^2\Big)^2+2\Big(\sum_{k=1}^n |a_k|^2\Big)\sum_{k=1}^n\sum_{l=k+1}^n (a_k\bar a_l+\bar a_k a_l)\\ &\quad+\sum_{i=1}^n\sum_{j=i+1}^n\sum_{k=1}^n\sum_{l=k+1}^n (a_i\bar a_j+\bar a_i a_j)(a_k\bar a_l+\bar a_k a_l)\\ &:= \Big(\sum_{k=1}^n |a_k|^2\Big)^2+A+B.\end{aligned}$$

Apply this in our case: $a_\ell = c_\ell e^{2\imath\pi\alpha S_\ell(\omega)}$, $i = 1, \dots, n$. The sum A equals to

$$A = 2\Big(\sum_{k=1}^n |c_k|^2\Big)\sum_{k=1}^n\sum_{l=k+1}^n \big(c_k\bar c_l e^{2\imath\pi\alpha(S_k(\omega)-S_l(\omega))}+\bar c_k c_l e^{2\imath\pi\alpha(S_l(\omega)-S_k(\omega))}\big).$$

By integrating over $\Omega\times\mathbf{T}$, with respect to $\mathbf{P}\times m$, we obtain an expression, which is equal to

$$\tilde A = 2\Big(\sum_{k=1}^n |c_k|^2\Big)\sum_{1\le k<l\le n}\big(c_k\bar c_l\mathbf{P}\{S_l = S_k\}+\bar c_k c_l\mathbf{P}\{S_l=-S_k\}\big).$$

Let

$$f_n(\alpha,\omega) = e^{2i\pi\alpha S_n(\omega)}.$$

We claim that

$$\sup_{n\ge1}\sum_{m\ge1}\big|\langle f_n, f_m\rangle_{\mathbf{P}\times\lambda}\big| \le 2\,G(0,0)$$

where $\langle\cdot,\cdot\rangle$ denotes covariance. Clearly,

$$0\le\langle f_n, f_m\rangle_{\mathbf{P}\times\lambda} = \mathbf{E}\int_{-1/2}^{1/2} e^{2i\pi\alpha(S_n-S_m)}d\alpha = \mathbf{P}\{S_n=S_m\} = \mathbf{P}\{S_{|m-n|}=0\}.$$

One one hand,

$$\sum_{m\ge n}\langle f_n, f_m\rangle_{\mathbf{P}\times\lambda} = \sum_{d\ge0}\mathbf{P}\{S_d=0\} = G(0,0),$$

and on the other,

$$\sum_{m<n}\langle f_n, f_m\rangle_{\mathbf{P}\times\lambda} = \sum_{d=1}^n\mathbf{P}\{S_d=0\} \le G(0,0).$$

Therefore

$$\sum_{m\ge1}\langle f_n, f_m\rangle_{\mathbf{P}\times\lambda} \le 2\,G(0,0).$$

Thus using Lemma 7.4.3 of Weber [48] it follows that $\{f_n, n \geq 1\}$ is a quasi-orthogonal system. As $\langle f_n, f_m\rangle_{\mathbf{P}\times\lambda} = \mathbf{P}\{S_{|m-n|} = 0\}$, it follows that

$$|\tilde{A}| \leq 4G(0,0)\Big(\sum_{k=1}^{n} |c_k|^2\Big). \tag{4.7}$$

Now, the sum B equals to

$$\sum_{1\leq i<j\leq n}\sum_{1\leq k<l\leq n} \big(\alpha_i\bar{\alpha}_j e^{2\imath\pi\alpha(S_i(\omega)-S_j(\omega))} + \bar{\alpha}_i\alpha_j e^{2\imath\pi\alpha(S_j(\omega)-S_i(\omega))}\big)$$
$$\times\big(\alpha_k\bar{\alpha}_l e^{2\imath\pi\alpha(S_k(\omega)-S_l(\omega))} + \bar{\alpha}_k\alpha_l e^{2\imath\pi\alpha(S_l(\omega)-S_k(\omega))}\big).$$

Integrating this expression over $\Omega\times\mathbf{T}$, with respect to $\mathbf{P}\times m$, we find a sum of the type

$$\tilde{B} = \sum_{1\leq i<j\leq n}\sum_{1\leq k<l\leq n} \gamma_i\,\gamma_j\,\gamma_k\,\gamma_l\,\mathbf{P}\big\{S_l - S_k = \pm(S_j - S_i)\big\},$$

where $\gamma_i = \alpha_i$ or $\bar{\alpha}_i$. Consider six cases.

i) $(1 \leq k < l \leq i)$ The sum differences $S_j - S_i$ and $S_l - S_k$ are independent, and we find in this case a contribution given by

$$\sum_{1\leq i<j\leq n}\sum_{1\leq k<l\leq i} \gamma_i\,\gamma_j\,\gamma_k\,\gamma_l\,\mathbf{P}\big\{S_j - S_i = \pm(S_l - S_k)\big\}.$$

ii) $(1 \leq k \leq i < l < j)$ There are two subcases: $S_l - S_k = S_j - S_i$ and $S_l - S_k = -(S_j - S_i)$. Write $a = i-k$, $b = l-i$, $c = j-l$. This corresponds to $a+b = \pm(b+c)$.

— if $a+b = b+c$, then $S_i - S_k = S_j - S_l$, which are independent sum differences. Hence a contribution equal to

$$\sum_{1\leq l<j\leq n}\sum_{1\leq k\leq i<l} \gamma_i\,\gamma_j\,\gamma_k\,\gamma_l\,\mathbf{P}\big\{S_i - S_k = S_j - S_l\big\}.$$

— if $a+b = -b-c$, then $a = -c-2b$ and $S_i - S_k = -(S_j - S_l) - 2(S_l - S_i)$, which are independent sum differences. Hence a contribution equal to

$$\sum_{1\leq l<j\leq n}\sum_{1\leq k\leq i\leq l} \gamma_i\,\gamma_j\,\gamma_k\,\gamma_l\,\mathbf{P}\big\{S_i - S_k = -(S_j - S_l) - 2(S_l - S_i)\big\}.$$

iii) $(1 \leq k \leq i < j \leq l \leq n)$ Write $a = i-k$, $b = j-i$, $c = l-j$. The equation $S_j - S_i = \pm(S_l - S_k)$ corresponds to $a+b+c = \pm b$.

— if $a+b+c = b$, then $a+c = 0$ and $(S_i - S_k) + (S_l - S_j) = 0$ where $S_i - S_k$ and $S_l - S_j$ are independent sum differences. Therefore, this produces a contribution equal to

$$\sum_{1\leq j\leq l\leq n}\sum_{1\leq k\leq i<j} \gamma_i\,\gamma_j\,\gamma_k\,\gamma_l\,\mathbf{P}\big\{(S_i - S_k) > S_l - S_j)\big\}.$$

— if $a+b+c = -b$, then $a = -c-2b$ or else $S_i - S_k = -(S_l - S_j) - 2(S_j - S_i)$ where $S_i - S_k$, $S_l - S_j$ and $S_j - S_i$ are independent sum differences. This produces a contribution equal to

$$\sum_{1\leq j\leq l\leq n}\sum_{1\leq k\leq i<j} \gamma_i\,\gamma_j\,\gamma_k\,\gamma_l\,\mathbf{P}\big\{S_i - S_k = -(S_l - S_j) - 2(S_j - S_i)\big\}.$$

iv) $(1 \le i < k < l \le j \le n)$ Write $a = k - i$, $b = l - k$, $c = j - l$. The equation $S_j - S_i = \pm(S_l - S_k)$ again corresponds to $a + b + c = \pm b$.

— if $a + b + c = b$, then $a + c = 0$ and $(S_k - S_i) + (S_j - S_l) = 0$ where $S_k - S_i$, $S_j - S_l$ are independent sum differences. This produces a contribution equal to

$$\sum_{1\le l\le j\le n}\ \sum_{1\le i<k<l} \gamma_i\,\gamma_j\,\gamma_k\,\gamma_l\,\mathbf{P}\big\{(S_k - S_i) + (S_j - S_l) = 0\big\}.$$

— if $a + b + c = -b$, then $a = -c - 2b$ and $S_k - S_i = -(S_j - S_l) - 2(S_l - S_k)$ where $S_k - S_i$, $S_j - S_l$ and $S_l - S_k$ are independent sum differences. This produces a contribution equal to

$$\sum_{1\le l\le j\le n}\ \sum_{1\le i<k<l} \gamma_i\,\gamma_j\,\gamma_k\,\gamma_l\,\mathbf{P}\big\{S_k - S_i = -(S_j - S_l) - 2(S_l - S_k)\big\}.$$

v) $(1 \le i < k < j < l \le n)$ Write $a = k - i$, $b = j - k$, $c = l - j$. The equation $S_j - S_i = \pm(S_l - S_k)$ corresponds here to $a + b = \pm(b + c)$.

— if $a + b = b + c$, then $a = c$ and $S_k - S_i = S_l - S_j$ where $S_k - S_i$, $S_l - S_j$ are independent sum differences. This produces a contribution equal to

$$\sum_{1\le j<l\le n}\ \sum_{1\le i<k\le j} \gamma_i\,\gamma_j\,\gamma_k\,\gamma_l\,\mathbf{P}\big\{S_k - S_i = S_l - S_j\big\}.$$

— if $a + b = -b - c$, then $a = -2b - c$ and $S_k - S_i = -(S_l - S_j) - 2(S_j - S_k)$ where $S_k - S_i$, $S_l - S_j$ and $S_j - S_k$ are independent sum differences. This produces a contribution equal to

$$\sum_{1\le j<l\le n}\ \sum_{1\le i<k\le j} \gamma_i\,\gamma_j\,\gamma_k\,\gamma_l\,\mathbf{P}\big\{S_k - S_i = -(S_l - S_j) - 2(S_j - S_k)\big\}.$$

vi) $(1 \le i < j \le k < l \le n)$ The sum differences $S_j - S_i$ and $S_l - S_k$ are independent, therefore in this case we find a contribution

$$\sum_{1\le i<j\le n}\ \sum_{1\le k<l\le i} \gamma_i\,\gamma_j\,\gamma_k\,\gamma_l\,\mathbf{P}\big\{S_j - S_i = \pm(S_l - S_k)\big\}.$$

Summarizing the above estimates, only two types of sums appear:

$$\sum_{1\le i\le k<l\le j\le n} \gamma_i\,\gamma_j\,\gamma_k\,\gamma_l\,\mathbf{P}\big\{S_k - S_i = \pm(S_j - S_l)\big\}. \tag{$S1$}$$

and

$$\sum_{1\le i\le k<l\le j\le n} \gamma_i\,\gamma_j\,\gamma_k\,\gamma_l\,\mathbf{P}\big\{S_k - S_i = (S_j - S_l) - 2(S_l - S_k)\big\} \tag{$S2$}$$

The proof is completed now by counting the number of occurrences of these sums, and using (4.7). □

We shall deduce from Proposition 4.1 a more explicit estimate of the fourth moment. We will use the following transform. Let $\gamma = \{\gamma_n, n \ge 1\}$ be a bounded sequence of non negative reals. Put for any $z \in \mathbf{Z}$

$$\gamma_h^{[z]} = \sum_{u\ge h} \gamma_u \mathbf{P}\{S_{u-h} = z\}.$$

By the transience assumption, these quantities are well defined since $\sum_{u\geq 0}\mathbf{P}\{S_u = z\} \leq G(0,0)$ for any $z \in \mathbf{Z}$. In particular, if γ is nonincreasing, we get from the above equality:

$$\gamma_h^{[z]} \leq G(0,0)\gamma_{z+h}.$$

In the case of a Bernoulli random walk, this is however read directly. As $\mathbf{P}\{S_{u-h}=z\} = 0$ if $z \leq 0$ or $z > u-h$, one has

$$\begin{aligned}\gamma_h^{[z]} &= \sum_{u\geq z+h} \gamma_u \mathbf{P}\{S_{u-h} = z\} \overset{(u=v+z+h)}{=} \sum_{v=0}^{\infty} \gamma_{v+z+h}\mathbf{P}\{S_{v+z} = z\} \\ &= \sum_{v=0}^{\infty} \gamma_{v+z+h} 2^{-(v+z)} C_{v+z}^z.\end{aligned}$$

Using now the formula $\sum_{v=0}^{\infty} C_{v+z}^z x^v = \frac{1}{(1-x)^{z+1}}$ valid for $|x| < 1$, gives the relation

$$\sum_{v=0}^{\infty} 2^{-(v+z)} C_{v+z}^z = 2$$

for any $z \geq 0$.

PROPOSITION 4.2. *Assume* $\mathbf{P}\{X \geq 0\} = 1$. *Let* $\mathbf{a} = \{a_k, k \geq 1\}$ *and* $\mathbf{c} = \{c_k, k \geq 1\}$ *be two sequences of reals such that* $|a_k| \leq c_k$ *for any* k *and* $\mathbf{c}$ *is nonincreasing. Then we have*

$$\begin{aligned}&\mathbf{E}\int_{\mathbf{T}} \Big|\sum_{k=m}^{n+m} a_k\, e^{2\pi\imath\alpha S_k}\Big|^4 d\alpha \\ &\leq 4G(0,0)\Big(\sum_{k=m}^{n+m} c_k^2\Big) + 48\Big\{\sum_{l=m}^{m+n} c_l^2\Big(\sum_{m\leq i\leq l} c_i\Big)^2 + \Big(\sum_{i=m}^{m+n} c_i\Big)^2 \sum_{l\geq m+n} c_l^2\Big\}.\end{aligned}$$

COROLLARY 4.1. *Assume* $\mathbf{P}(X \geq 0) = 1$. *Let* $\mathbf{a} = \{a_k, k \geq 1\}$ *and* $\mathbf{c} = \{c_k, k \geq 1\}$ *be two sequences of reals such that* $|a_k| \leq c_k$ *for any* k *and* $\mathbf{c}$ *is nonincreasing. Then the series* $\sum_{k=1}^{\infty} a_k\, e^{2\pi\imath\alpha S_k}$ *converges in* $L^4(\mathbf{P}\times\lambda)$, *provided that the series*

$$\sum_{l\geq 1} c_l^2 \Big(\sum_{1\leq i\leq l} c_i\Big)^2$$

converges. In particular the series $\sum_{k=1}^{\infty} k^{-a}\, e^{2\imath\pi\alpha S_k}$ *converges in* $L^4(\mathbf{P}\times\lambda)$ *for any* $a > 3/4$.

PROOF OF PROPOSITION 4.2. By Proposition 4.1

$$\mathbf{E}\int_{\mathbf{T}} \Big|\sum_{k=m}^{n+m} a_k\, e^{2\imath\pi\alpha S_k}\Big|^4 d\alpha \leq 4G(0,0)\Big(\sum_{k=m}^{n+m} c_k^2\Big) + 6((S1) + (S2)),$$

where

$$\begin{aligned}(S1) &= \sum_{m\leq i\leq k<l\leq j\leq m+n} c_i c_j c_k c_l \mathbf{P}\big\{S_k - S_i = \pm(S_j - S_l)\big\}, \\ (S2) &= \sum_{m\leq i\leq k<l\leq j\leq m+n} c_i c_j c_k c_l \mathbf{P}\big\{S_k - S_i = (S_j - S_l) - 2(S_l - S_k)\big\}.\end{aligned}$$

Consider first the sums of type $(S1)$, the others will be in turn treated similarly. Write

$$\begin{aligned}\sum_{m\le i\le k<l\le j\le m+n} & c_ic_jc_kc_l\mathbf{P}\{S_k-S_i=S_j-S_l\}\\ &=\sum_{z\in\mathbf{Z}}\sum_{m\le i\le k\le m+n} c_ic_k\mathbf{P}\{S_k-S_i=z\}\sum_{k<l\le j\le m+n} c_jc_l\mathbf{P}\{S_j-S_l=z\}\\ &=\sum_{z\in\mathbf{Z}}\sum_{m\le i\le k\le m+n} c_ic_k\mathbf{P}\{S_{k-i}=z\}\sum_{k<l\le j\le m+n} c_jc_l\mathbf{P}\{S_{j-l}=z\}.\end{aligned}$$

As

$$\sum_{k<l\le j\le m+n} c_jc_l\,\mathbf{P}\{S_{j-l}=z\}\le\sum_{m\le l\le m+n} c_l\Big(\sum_{j\ge l} c_j\mathbf{P}\{S_{j-l}=z\}\Big)=\sum_{m\le l\le m+n} c_lc_l^{[z]},$$

we get by putting this into the previous relation

$$\begin{aligned}&\sum_{m\le i\le k<l\le j\le m+n} c_ic_jc_kc_l\,\mathbf{P}\{S_k-S_i=S_j-S_l\}\le\\ &\sum_{z\in\mathbf{Z}}\Big(\sum_{l=m}^{m+n} c_lc_l^{[z]}\Big)\sum_{m\le i\le m+n} c_i\Big(\sum_{k\ge i} c_k\mathbf{P}\{S_{k-i}=z\}\Big)\le\sum_{z\in\mathbf{Z}}\Big(\sum_{l=m}^{m+n} c_lc_l^{[z]}\Big)\sum_{i=m}^{m+n} c_ic_i^{[z]}\\ &\qquad\le\sum_{m\le i,l\le m+n} c_lc_i\sum_{z\in\mathbf{Z}} c_l^{[z]}c_i^{[z]}.\end{aligned}$$

The sums related to the factor $\mathbf{P}\{S_k-S_i=-(S_j-S_l)\}$ are treated similarly; the latter probability being not 0 only if $\mathbf{P}\{S_k-S_i=0\}=\mathbf{P}\{S_j-S_l=0\}$, and its value is then $\mathbf{P}\{S_k-S_i=0\}\mathbf{P}\{S_j-S_l=0\}$. Thus

$$\begin{aligned}&\sum_{m\le i\le k<l\le j\le m+n} c_ic_jc_kc_l\mathbf{P}\{S_k-S_i=-(S_j-S_l)\}\\ &=\sum_{m\le i\le k\le m+n} c_ic_k\mathbf{P}\{S_k-S_i=0\}\sum_{k<l\le j\le m+n} c_lc_j\mathbf{P}\{S_j-S_l=0\}\\ &\le\sum_{m\le i\le k\le m+n} c_ic_k\mathbf{P}\{S_k-S_i=0\}\sum_{k<l\le m+n} c_l\sum_{j\ge l} c_j\mathbf{P}\{S_j-S_l=0\}\\ &\le\sum_{m\le i\le k\le m+n} c_ic_k\mathbf{P}\{S_k-S_i=0\}\Big(\sum_{m\le l\le m+n} c_lc_l^{[0]}\Big)\\ &\le\sum_{i=m}^{m+n} c_i\Big(\sum_{k\ge i} c_k\mathbf{P}\{S_k-S_i=0\}\Big)\Big(\sum_{l=m}^{m+n} c_lc_l^{[0]}\Big)\le\sum_{i=m}^{m+n} c_ic_i^{[0]}\Big(\sum_{l=m}^{m+n} c_lc_l^{[0]}\Big)\\ &\le\Big(\sum_{i=m}^{m+n} c_i^2\Big)^2.\end{aligned}$$

Now, consider the sums of type $(S2)$:

$$\sum_{m\le i\le k<l\le j\le m+n} c_i c_j c_k c_l\, \mathbf{P}\big\{S_k - S_i = (S_j - S_l) - 2(S_l - S_k)\big\}$$
$$= \sum_{\substack{z_1\in\mathbf{Z}\\ z_2\in\mathbf{Z}}} \sum_{m\le i\le k<l\le j\le m+n} c_i\, c_j\, c_k\, c_l \mathbf{P}\big\{S_{k-i} = z_1\big\}\mathbf{P}\big\{S_{l-k} = z_2\big\}\mathbf{P}\big\{S_{j-l} = z_1 + 2z_2\big\}$$
$$\le \sum_{\substack{z_1\in\mathbf{Z}\\ z_2\in\mathbf{Z}}} \sum_{i=m}^{n+m} |c_i|$$
$$\times \left\{ \sum_{k=i}^{m+n} c_k \mathbf{P}\big\{S_{k-i} = z_1\big\} \sum_{l=k}^{m+n} c_l \Big(\sum_{j\ge l} |c_j| \mathbf{P}\big\{S_{j-l} = z_1 + 2z_2\big\}\Big) \mathbf{P}\big\{S_{l-k} = z_2\big\}\right\}$$
$$\le \sum_{\substack{z_1\in\mathbf{Z}\\ z_2\in\mathbf{Z}}} \sum_{i=m}^{n+m} c_i \Big\{ \sum_{i\le k\le n+m} c_k \mathbf{P}\big\{S_{k-i} = z_1\big\} \sum_{k\le l\le n+m} c_l c_l^{[z_1+2z_2]} \mathbf{P}\big\{S_{l-k} = z_2\big\}\Big\}.$$

Consider on $L^2(\mathbf{T})$, the operator U defined for $h \sim \sum_{z\in\mathbf{N}} h_z e_z$ by

$$Uh \sim \sum_{z\in\mathbf{N}} h_{z+1} e_z.$$

Let $g = \sum_{k=1}^{\infty} a_k e_k$. It follows that

$$\sum_{i,l=m}^{n+m} c_l c_i \sum_{z\in\mathbf{N}} c_l^{[z]} c_i^{[z]} \le 4 \sum_{i,l=m}^{n+m} \sum_{z\in\mathbf{N}} c_l c_i c_{l+z} c_{i+z}$$
$$= 4 \sum_{i,l=m}^{n+m} c_l c_i \langle U^l g, U^i g\rangle = 4 \Big\| \sum_{i=m}^{n+m} c_i U^i g\Big\|^2$$

and

$$\sum_{\substack{z_1\in\mathbf{Z}\\ z_2\in\mathbf{Z}}} \sum_{i=m}^{n+m} c_i \Big\{ \sum_{i\le k\le m+n} c_k \mathbf{P}\big\{S_{k-i} = z_1\big\} \sum_{k<l\le m+n} c_l c_l^{[z_1+2z_2]} \mathbf{P}\big\{S_{l-k} = z_2\big\}\Big\}$$
$$\le 2 \sum_{\substack{z_1\in\mathbf{Z}\\ z_2\in\mathbf{Z}}} \sum_{i=m}^{n+m} c_i \Big\{ \sum_{i\le k\le m+n} c_k \mathbf{P}\big\{S_{k-i}=z_1\big\} \sum_{k<l\le m+n} c_l c_{l+z_1+2z_2} \mathbf{P}\big\{S_{l-k}=z_2\big\}\Big\}$$
$$\le 2 \sum_{\substack{z_1\in\mathbf{Z}\\ z_2\in\mathbf{Z}}} \sum_{i=m}^{n+m} c_i \Big\{ \sum_{i\le k\le m+n} c_k \mathbf{P}\big\{S_{k-i} = z_1\big\} \sum_{l\ge k+z_2} c_l c_{l+z_1+2z_2} \mathbf{P}\big\{S_{l-k} = z_2\big\}\Big\}$$
$$\le 4 \sum_{\substack{z_1\in\mathbf{N}\\ z_2\in\mathbf{N}}} \sum_{i=m}^{n+m} c_i \Big\{ \sum_{i+z_1\le k\le m+n} c_k \mathbf{P}\big\{S_{k-i} = z_1\big\} c_{k+z_2} c_{k+z_1+3z_2}\Big\}$$
$$\le 4 \sum_{i=m}^{n+m} c_i \sum_{k=m}^{n+m} c_k \Big\{ \sum_{z_1\le k-i} \mathbf{P}\big\{S_{k-i} = z_1\big\}\Big\} \sum_{z_2\in\mathbf{N}} c_{k+z_2} c_{i+z_2}$$
$$= 4 \sum_{i,k=m}^{n+m} c_i c_k \langle U^i g, U^k g\rangle = 4 \Big\| \sum_{m\le i\le m+n} c_i U^i g\Big\|^2.$$

Consequently,

$$\mathbf{E}\int_{\mathbf{T}}\Big|\sum_{k=m}^{n+m} a_k\, e^{2\imath\pi\alpha S_k}\Big|^4 d\alpha \le 4G(0,0)\Big(\sum_{k=m}^{n+m} c_k^2\Big) + 48\Big\{\Big(\sum_{k=m}^{n+m} c_k^2\Big)^2 + \Big\|\sum_{m\le i\le m+n} c_i U^i g\Big\|^2\Big\}.$$

As

$$\sum_{i=m}^{m+n} c_i U^i g = \sum_{i=m}^{m+n}\sum_{l\ge i} c_i c_l e_l = \sum_{l\ge m} e_l c_l\Big(\sum_{i=m}^{(m+n)\wedge l} c_i\Big)$$

$$= \sum_{l=m}^{m+n} e_l c_l\Big(\sum_{m\le i\le l} c_i\Big) + \Big(\sum_{m\le i\le m+n} c_i\Big)\sum_{l\ge m+n} e_l c_l,$$

one has

$$\Big\|\sum_{i=m}^{m+n} c_i U^i g\Big\|^2 = \sum_{l=m}^{m+n} c_l^2\Big(\sum_{m\le i\le l} c_i\Big)^2 + \Big(\sum_{i=m}^{m+n} c_i\Big)^2 \sum_{l\ge m+n} c_l^2.$$

Hence

$$\mathbf{E}\int_{\mathbf{T}}\Big|\sum_{k=m}^{n+m} a_k\, e^{2\imath\pi\alpha S_k}\Big|^4 d\alpha$$

$$\le 4G(0,0)\Big(\sum_{k=m}^{n+m} c_k^2\Big) + 48\Big\{\sum_{l=m}^{m+n} c_l^2\Big(\sum_{m\le i\le l} c_i\Big)^2 + \Big(\sum_{i=m}^{m+n} c_i\Big)^2\sum_{l\ge m+n} c_l^2\Big\}.$$

□

We turn now to the study of convergence of $\sum_{k=1}^{\infty} c_k f(S_k x)$ for general $f \in L^2(\mathbf{T})$, $\int_{\mathbf{T}} f(t)dt = 0$. In the case when the distribution of X_1 is absolutely continuous, the exact analogue of Theorem 4.1 holds, namely we have

THEOREM 4.2. *Let X_1 have a bounded density. Let further $f \in \mathrm{Lip}_\alpha(\mathbf{T})$ for some $\alpha > 0$ with $\int_{\mathbf{T}} f(t)dt = 0$ and let $\mathbf{c} \in \ell^2$. Then for any fixed $x \neq 0$, $\sum_{k=1}^{\infty} c_k f(S_k x)$ converges with probability 1. Consequently, for almost every $\omega \in \Omega$, $\sum_{k=1}^{\infty} c_k f(S_k(\omega)x)$ converges for almost every x.*

In the case when X_1 has a lattice distribution, the situation is more complicated. We will prove the following result.

THEOREM 4.3. *Let $X_1 \ge 0$ be an integer valued random variable with finite expectation and $\mathbf{P}(X_1 = 0) < 1$. Let $f \in L^2(\mathbf{T})$ with $\int_{\mathbf{T}} f(t)dt = 0$ have Fourier series*

$$f \sim \sum_{k=1}^{\infty}(a_k \cos 2\pi kx + b_k \sin 2\pi kx)$$

and assume that the Dirichlet series $\sum_n a_n n^{-s}$, $\sum_n b_n n^{-s}$ are regular and bounded in the half-plane $\Re(s) > 0$. Put

$$\tau_{k,\gamma_1,\gamma_2}(\mathbf{c}) := \sup_{\substack{L\ge\gamma_1 k\\ u\le\gamma_2\log k}} \sum_{\ell=L}^{L+u} |c_\ell|.$$

Then the series $\sum_{k=1}^{\infty} c_k f(S_k(\omega)x)$ converges in $L^2(\mathbf{T})$ norm for $\mathbf{P}$-almost every ω provided $\mathbf{c} \in \ell^2$ and $\tau_{k,\gamma_1,\gamma_2}(\mathbf{c}) = o(1)$ $(k \to \infty)$ for any $\gamma_1, \gamma_2 > 0$.

For the proof of Theorem 4.2 we change the notation slightly and let S_n denote the fractional part of $X_1 + \ldots + X_n$. As we only deal with the series $\sum c_k f(S_k x)$ with f having period 1, this does not cause any problem. Put

$$\psi(x) = \sup_{0 \le x \le 1} |\mathbf{P}(S_n \le x) - x|$$

and note that by Theorem 1 of Schatte [43] it follows easily

$$\psi(n) \le Ce^{-\lambda n} \qquad (n \ge 1) \tag{4.8}$$

and for any $f \in \text{Lip}_\alpha(\mathbf{T})$, $\alpha > 0$

$$\left| \mathbf{E} f(S_n) - \int_0^1 f(x)dx \right| \le Ce^{-\lambda n} \qquad (n \ge 1) \tag{4.9}$$

for some constants $C > 0$, $\lambda > 0$, depending on f.

LEMMA 4.3. *Let $k < \ell < m < n$ be positive integers. Then there exists a r.v. Δ with $|\Delta| \le \psi(\ell - k)$ such that the vector $(S_\ell - \Delta, S_m - \Delta, S_n - \Delta)$ has uniform coordinates and is independent of $S_k - \Delta$. Similarly, there exists a random variable Δ' with $|\Delta'| \le \psi(n - m)$ such that $S_n - \Delta'$ has uniform coordinates and is independent of the vector $(S_k - \Delta', S_\ell - \Delta', S_m - \Delta')$.*

This lemma is implicit in Schatte [44] and can be obtained along the following lines. By enlarging the underlying probability space if necessary, we can assume that there exists a random variable U, uniformly distributed over $(0,1)$, and independent of the sequence $X_1, X_2, \ldots$ Let $Y = S_\ell - S_k$, then $|P(Y \le t) - t| \le \psi(\ell - k)$ for all t and thus by Lemma 3 of Schatte [44] there exists a uniform r.v. Y^*, which is a function of U and Y such that $|Y - Y^*| \le \psi(\ell - k)$. Let $\Delta = Y - Y^*$, then

$$(S_\ell - \Delta, S_m - \Delta, S_n - \Delta) = (S_\ell - Y, S_m - Y, S_n - Y) + Y^* =: Z + Y^*.$$

Here the vector Z is obviously independent of $Y = X_{k+1} + \cdots + X_\ell$ and thus also of the uniform r.v. Y^*, which is a function of Y and U. Thus adding Y^* to the components of Z, we get a vector whose components are uniform (see Lemma 1 of [44]), the independence of $Z + Y^*$ and S_k follows also from Lemma 1 of [44]. The second statement of the lemma can be proved similarly.

PROOF OF THEOREM 4.2. We prove the statement for $x = 1$. Let $f \in \text{Lip}_\alpha(\mathbf{T})$, $\alpha > 0$ with $\int_{\mathbf{T}} f(t)dt = 0$. Set $\xi_k = f(S_k) - \mathbf{E} f(S_k)$. By (4.9), for any bounded sequence (c_k) the series $\sum c_k f(S_k)$ and $\sum c_k \xi_k$ are equiconvergent, and thus it suffices to prove that $\sum c_k \xi_k$ is a.s. convergent provided $(c_k) \in \ell_2$. In view of Lemma 4.1, this will follow if we show that

$$\mathbf{E}\Bigg(\sum_{k=1}^N c_k \xi_k \Bigg)^4 \le K \Bigg(\sum_{k=1}^N c_k^2 \Bigg)^2 \tag{4.10}$$

for any real $(c_k)_{k=1}^N$ with a suitable constant K. We claim that

$$|\mathbf{E}(\xi_k \xi_l \xi_m \xi_n)| \le Ae^{-C(|l-k|+|n-m|)} \qquad (k \le l \le m \le n). \tag{4.11}$$

By Lemma 4.3, there exists a r.v. Δ with $|\Delta| \le \psi(l - k)$ such that the vector

$$(S_l - \Delta, S_m - \Delta, S_n - \Delta) =: (S_l', S_m', S_n')$$

is independent of S_k and thus the r.v.'s

$$X = f(S_k) - \mathbf{E}f(S_k) \text{ and } Y = (f(S_l') - \mathbf{E}f(S_l))(f(S_m') - \mathbf{E}f(S_m))(f(S_n') - \mathbf{E}f(S_n))$$

are independent. Since $\mathbf{E}(X) = 0$, it follows that

$$\mathbf{E}\big(f(S_k) - \mathbf{E}f(S_k)\big)\big(f(S_l') - \mathbf{E}f(S_l)\big)\big(\big(f(S_m') - \mathbf{E}f(S_m)\big)\big(f(S_n') - \mathbf{E}f(S_n)\big)\big)$$
$$= \mathbf{E}(XY) = \mathbf{E}(X)\mathbf{E}(Y) = 0. \qquad (4.12)$$

In view of $|\Delta| \le \psi(l-k)$ and the boundedness and Lipschitz property of f it follows that replacing S_l', S_m', S_n' by S_l, S_m, S_n in the first expectation in (4.12) results in a change of at most $C\psi(l-k)$ of the expectation and thus we see that the expectation in (4.11) is at most $C\psi(l-k)$. A similar argument shows (using the second statement of Lemma 4.3) that the left-hand side of (4.11) is at most $C\psi(n-m)$, and thus the left-hand side of (4.11) is also bounded by $C(\psi(l-k)\psi(n-m))^{1/2}$, which proves (4.11) in view of (4.8).

It is now easy to verify (4.10). By (4.11), the left-hand side of (4.10) is bounded by

$$24 \sum_{1\le k\le l\le m\le n\le N} |c_k||c_l||c_m||c_n|e^{-C(|l-k|+|n-m|)}. \qquad (4.13)$$

The right-hand side here is the same as that of (4.6) and repeating the argument at the end of the proof of Lemma 4.2 we get (4.10). □

Proof of Theorem 4.3. Let $I_k = [\Delta_k, \Delta_k^*]$, $k = 0, 1, \dots$ denote the (random) interval of n's for which $S_n = k$. Some of these intervals can be empty and the union of the nonempty ones is $\mathbf{N}$. Let

$$Y_k = \sum_{\ell\in[\Delta_k,\Delta_k^*]} c_\ell.$$

Let $\mathcal{G}$ be the set of k's ("good indices") for which I_k is nonempty. Purely formally,

$$\sum_{k=1}^{\infty} c_k f(S_k x) = \sum_{h=0}^{\infty} Y_h f(hx)$$

and to prove the convergence of the sum on the left-hand side we first prove the convergence of the right-hand side. To this end, we first show

Lemma 4.4. *For any* $\mathbf{c} \in \ell^2$, *the series* $\sum_{k=0}^{\infty} \mathbf{E}Y_k^2$ *converges.*

Proof. Clearly

$$Y_k = \sum_{\ell=1}^{\infty} c_\ell I\{\Delta_k \le \ell \le \Delta_k^*\}$$

and thus

$$\mathbf{E}Y_k^2 = \mathbf{E}\left|\sum_{i,j=1}^{\infty} c_i c_j I\{\Delta_k \le i, j \le \Delta_k^*\}\right| \le \sum_{i,j=1}^{\infty} |c_i||c_j|\mathbf{P}\{\Delta_k \le i, j \le \Delta_k^*\}.$$

Summing for k we get

$$\sum_{k=1}^{\infty} \mathbf{E}Y_k^2 \le \sum_{i,j=1}^{\infty} |c_i||c_j|\mathbf{P}\{i \text{ and } j \text{ belong to the same interval } [\Delta_k, \Delta_k^*]\}$$
$$=: \sum_{i,j=1}^{\infty} |c_i||c_j|\mathbf{P}(A_{i,j}).$$

Letting $\rho = \mathbf{P}(X_1 = 0) < 1$, we have for $i \le j$

$$\mathbf{P}(A_{i,j}) = P(S_i = S_j) = \mathbf{P}(X_{i+1} = \ldots = X_j = 0) = \rho^{j-i}$$

and thus

$$\sum_{k=1}^{\infty} \mathbf{E}Y_k^2 \le \sum_{i,j=1}^{\infty} |c_i||c_j|\rho^{|i-j|} \le 2\sum_{m=0}^{\infty} \rho^m \sum_{i=1}^{\infty} |c_i||c_{i+m}| \le 2\sum_{m=1}^{\infty} \rho^m \left(\sum_{i=1}^{\infty} c_i^2\right) < \infty,$$

completing the proof.

From Lemma 4.4 it follows that if $\mathbf{c} \in \ell^2$, then $\sum_{k=1}^{\infty} Y_k^2 < \infty$ almost surely, and thus applying Theorem A in Chapter 1 it follows that with $\mathbf{P}$-probability 1, the series $\sum_{h=0}^{\infty} Y_h f(hx)$ converges in $L^2(\mathbf{T})$ norm. Since

$$\sum_{\ell \le \Delta_k^*} c_\ell f(S_\ell x) = \sum_{h=0}^{k} Y_h f(hx) \qquad k \in \mathcal{G}$$

this means that with $\mathbf{P}$-probability 1 the partial sums $\sum_{k \le N} c_k f(S_k x)$ converge in $L^2(\mathbf{T})$ norm along the indices $N = \Delta_k^*$, $k \in \mathcal{G}$. It remains to show that almost surely

$$\max_{\Delta_k \le L \le \Delta_k^*} \int_{\mathbf{T}} \left| \sum_{L \le \ell \le \Delta_k^*} c_\ell f(S_\ell x) \right|^2 dx \to 0 \quad \text{as } k \to \infty,\ k \in \mathcal{G}. \tag{4.14}$$

We note that almost surely for $k \in \mathcal{G}$

$$\Delta_k \ge \gamma_1 k, \qquad \Delta_k^* - \Delta_k \le \gamma_2 \log k \qquad \text{ultimately}$$

for some positive constants γ_1, γ_2. The first relation is immediate from $0 < \mathbf{E}X_1 < \infty$ and the strong law of large numbers, while the second relation follows from $\mathbf{P}(\Delta_k^* - \Delta_k = m) = \rho^m$ $(m \in \mathcal{G})$ and the Borel–Cantelli lemma. Since S_n takes the constant value k on the interval $[\Delta_k, \Delta_k^*]$, the left-hand side of (4.14) is bounded by $\tau_{k,\gamma_1,\gamma_2}^2(\mathbf{c})\|f\|_2^2$, proving (4.14) and thus completing the proof of Theorem 4.3.□

CHAPTER 5

Discrepancy of random sequences $\{S_n x\}$

Given a sequence $\mathbf{s} = \{s_n, n \geq 1\}$ of real numbers in $[0,1)$, the discrepancy $D_N(\mathbf{s})$ of $\mathbf{s}$ is defined by

$$D_N(\mathbf{s}) = \sup_{0 \leq a < b \leq 1} \frac{1}{N} \Bigg| \sum_{\substack{n=1 \\ s_n \in [a,b)}}^{N} 1 - N(b-a) \Bigg|. \tag{5.1}$$

Clearly, $D_N(\mathbf{s})$ measures how far the distribution of $\mathbf{s}$ is from the uniform. The sequence $\mathbf{s}$ is called uniformly distributed in the Weyl sense if $D_N(\mathbf{s}) \to 0$ as $N \to \infty$. In this section we study the discrepancy $D_N(x, \omega)$ of the random sequence

$$\{S_n x,\ n \geq 1\} = \{S_n(\omega)x,\ n \geq 1\} \pmod 1$$

where $S_n = \sum_{k=1}^n X_k$ and X_k are i.i.d. random variables defined on some probability space $(\Omega, \mathcal{A}, \mathbf{P})$. Letting $f_{a,b}(t) = I_{a,b}(t) - (b-a)$ extended with period 1, we can write $D_n(x, \omega)$ as

$$D_N(x, \omega) = \frac{1}{N} \sup_{0 \leq a < b \leq 1} \Bigg| \sum_{k=1}^{N} f_{a,b}(S_k(\omega)x) \Bigg|$$

which is closely related to the convergence problems studied in Section 4.

In the case when the distribution of X_1 is absolutely continuous, the behavior of D_N is well known and is similar to the discrepancy of i.i.d. random variables. In fact Schatte [45] proved the following result:

THEOREM E. *Let $X_1, X_2, \ldots$ be i.i.d. random variables from the interval $[0,1)$, defined on some probability space $(\Omega, \mathcal{A}, \mathbf{P})$. Assume that the distribution of X_1 is absolutely continuous and put $S_n = X_1 + \cdots + X_n$. Let $D_N(x, \omega)$ denote the discrepancy of $\{S_k(\omega)x\}_{1 \leq k \leq N}$ mod 1. Then for every x and almost every ω we have*

$$0 < \limsup_{N \to \infty} \sqrt{\frac{N}{\log \log N}} D_N(x, \omega) < +\infty.$$

The purpose of this chapter is to study the case when the X_n have a lattice distribution. In this case, the Diophantine approximation properties of x play a crucial role, similarly to the case of nonrandom X_n. Previous results of this type were proved in Weber [50]. We first prove the following result:

THEOREM 5.1. *Let $X_1, X_2, \ldots$ be a sequence of independent, identically distributed lattice random variables defined on some probability space $(\Omega, \mathcal{A}, \mathbf{P})$. We assume that the random walk $S_n = X_1 + \ldots + X_n$, $n \geq 1$ is transient. Let $D_N(x, \omega)$ denote the discrepancy of $\{S_k(\omega)x\}_{1 \leq k \leq N}$ mod 1. Then, for any $\tau > 5/2$,*

$$D_N(x, \omega) \overset{a.s.}{=} \mathcal{O}(N^{-1/2} \log^\tau N) \tag{5.2}$$

for every $x \in \mathbf{R}$ and almost every ω.

To clarify the meaning of Theorem 5.1, recall that by classical results of Cassels [9] and Erdős and Koksma [11], for any increasing sequence (n_k) of positive integers, the discrepancy of $\{n_k x\}$ is $O(N^{-1/2}\log^\tau N)$ for almost every x and for any $\tau > 5/2$. Of course, this implies Theorem 5.1 in the case $X > 0$ a.s. In the general transient case, n_k can be negative and $|n_k|$ is not necessarily increasing, but we have $|n_k| \to \infty$ a.s. and thus with probability one, every term of (n_k) is repeated only finitely many times. This is a situation similar to that in the results of Cassels [9] and Erdős and Koksma [11], but one should observe that repetitions in a sequence of real numbers can change the discrepancy of the sequence drastically, even if we permit only finitely many repetitions of each term. The heuristic meaning of Theorem 5.1 is that repetitions in the sequence S_n are sufficiently limited so that the order of magnitude of the discrepancy of $\{S_n x\}$ remains the same as in the strictly monotone case.

It is worth mentioning that the constant 5/2 in the theorems of Erdős, Cassels and Koksma has been improved to 3/2 by R. C. Baker [1]. Of course, this raises the question if Theorem 5.1 also holds with $\tau > 3/2$ instead of $\tau > 5/2$. In the remark after the proof of Theorem 5.2 we will show that the constant 5/2 can be improved to 7/4 if the characteristic function φ of X satisfies $|\varphi(t) - 1| \geq C|t|$ for $|t| \leq t_0$; this is satisfied e.g. if X has a finite, nonzero mean. The argument there can be easily generalized for other classes of random variables X. Whether Theorem 5.1 holds with $\tau > 3/2$ for all transient X remains open.

For the proof of Theorem 5.1, we need some lemmas. Put for any integers $N \geq 1$, $m \geq 0$,

$$\Theta_N(m, x) = \sum_{n=1}^{N} e^{2i\pi m S_n x}. \tag{5.3}$$

LEMMA 5.1. *For any two integers $N \geq P \geq 1$, one has the following estimate:*

$$\mathbf{E}\int_{\mathbf{T}} \left|\Theta_N(m, x) - \Theta_P(m, x)\right|^2 dx \leq C_{\mathcal{X}}(N - P),$$

where the constant $C_{\mathcal{X}}$ depends on $\mathcal{X}$ only.

PROOF. Since

$$\begin{aligned}
&\mathbf{E}\int_{\mathbf{T}} \left|\Theta_N(m, x) - \Theta_P(m, x)\right|^2 dx \\
&= \mathbf{E}\int_{\mathbf{T}} \sum_{P<k,\ell\leq N} e^{2i\pi m(S_k - S_\ell)x} dx = \sum_{P<k,\ell\leq N} \mathbf{P}\{S_k = S_\ell\},
\end{aligned}$$

and

$$\begin{aligned}
&\sum_{P<k,\ell\leq N} \mathbf{P}\{S_k = S_\ell\} \\
&= (N - P) + 2\sum_{P<k<\ell\leq N} \mathbf{P}\{S_{\ell-k} = 0\} \leq (N - P)\Big\{1 + 2\Big(\sum_{\lambda\geq 1} \mathbf{P}\{S_\lambda = 0\}\Big)\Big\}.
\end{aligned}$$

The last sum is convergent by the transience of the random walk $\{S_n,\, n \geq 1\}$ and Lemma 5.1 follows. □

Put for any positive integer n and $x \in \mathbf{T}$,

$$U_n(x) = \sum_{h=1}^{n} \frac{1}{h} \big|\Theta_n(h,x)\big|. \tag{5.4}$$

LEMMA 5.2. *For any two integers $n > \ell \geq 1$,*

$$\mathbf{E} \int_{\mathbf{T}} \big|U_n(x) - U_\ell(x)\big|^2 dx \leq C_{\mathcal{X}} \{(n-\ell)\log^2 \ell + n \log^2(n/l)\}. \tag{5.5}$$

PROOF. Clearly,

$$U_\ell(x) - U_n(x) = \sum_{h=1}^{\ell} \frac{1}{h}\big(|\Theta_\ell(h,x)| - |\Theta_n(h,x)|\big) - \sum_{h=\ell+1}^{n} \frac{1}{h}|\Theta_n(h,x)| := A - B.$$

By the Cauchy–Schwarz inequality and Lemma 5.1,

$$\mathbf{E} \int_{\mathbf{T}} A^2 d\lambda \leq \Big(\sum_{h=1}^{\ell} \frac{1}{h}\Big)\Big(\sum_{h=1}^{\ell} \frac{1}{h} \mathbf{E} \int_{\mathbf{T}} \big|\Theta_\ell(h,x) - \Theta_n(h,x)\big|^2 dx\Big) \leq C_{\mathcal{X}}(n-\ell)\log^2 \ell,$$

$$\mathbf{E} \int_{\mathbf{T}} B^2 d\lambda \leq \Big(\sum_{h=\ell+1}^{n} \frac{1}{h}\Big)\Big(\sum_{h=\ell+1}^{n} \frac{1}{h} \mathbf{E} \int_{\mathbf{T}} \big|\Theta_n(h,x)\big|^2 dx\Big) \leq C_{\mathcal{X}} n \log^2(n/\ell).$$

Lemma 5.2 thus follows. □

LEMMA 5.3. *For any $\tau > 5/2$,*

$$U_n \overset{a.s.}{=} \mathcal{O}\big(n^{1/2} \log^\tau n\big). \tag{5.6}$$

PROOF. By the concavity of $\log x$ we have for any $n > \ell \geq 1$

$$\frac{\log n - \log \ell}{n - \ell} \leq \frac{\log n}{n}.$$

Thus by Lemma 5.2

$$\begin{aligned} &\mathbf{E} \int_{\mathbf{T}} \big|U_n(x) - U_\ell(x)\big|^2 dx \\ &\leq C_{\mathcal{X}} \log n \big\{(n-\ell)\log \ell + n \log(n/\ell)\big\} \leq C_{\mathcal{X}}(n-\ell)\log^2 n. \end{aligned} \tag{5.7}$$

Hence,

$$\mathbf{E} \int_{\mathbf{T}} \big|U_n(x) - U_\ell(x)\big|^2 dx \leq C_{\mathcal{X}}(n-\ell)\log^2 n, \qquad \mathbf{E} \int_{\mathbf{T}} \big|U_n(x)\big|^2 dx \leq C_{\mathcal{X}} n \log^2 n.$$

Let $a > 1/2$. By the Chebysev inequality,

$$\mathbf{P} \times \lambda\big\{|U_{2^p}| > (2^p\, p^2)^{1/2} p^a\big\} \leq C_{\mathcal{X}} p^{-2a},$$

and thus the Borel–Cantelli Lemma yields

$$|U_{2^p}| \overset{a.s.}{=} \mathcal{O}\big((2^p p^2)^{1/2} p^a\big)$$

Now, investigate the oscillation of U_n over the interval $[2^p, 2^{p+1})$. Put

$$U_n' = U_n/(2^p p^2)^{1/2}.$$

Then

$$\mathbf{E}\big|U'_n - U'_\ell\big|^2 \le C\Big(\frac{n-\ell}{2^p}\Big).$$

Applying Lemma 3.4 of Weber [49], gives

$$\Big\| \sup_{2^p \le n,m < 2^{p+1}} \big|U'_n - U'_\ell\big| \Big\|_{2,\mathbf{P}\times\lambda} \le C_{\mathcal{X}} p.$$

Let $\beta > 3/2$. By the Tchebycheff inequality,

$$\mathbf{P}\Big\{ \sup_{2^p \le n,m < 2^{p+1}} \big|U_n - U_l\big| > (2^p p^2)^{1/2} p^\beta \Big\} \le C\, p^{2-2\beta},$$

which implies by the Borel–Cantelli Lemma

$$\sup_{2^p \le n,m < 2^{p+1}} \big|U_n - U_l\big| \overset{a.s.}{=} \mathcal{O}\big((2^p p^2)^{1/2} p^\beta\big)$$

Combining our estimates easily gives the result.

By the Erdős–Turán inequality (see e.g. Harman [22], Theorem 5.5, p. 129) we have for any sequence $\mathbf{s} = \{s_n,\ n \ge 1\}$ and any positive integers L and N

$$ND_N(\mathbf{s}) \le \frac{N}{L+1} + C\sum_{h=1}^{L} \frac{1}{h}\Big|\sum_{n=1}^{N} e^{2i\pi h s_n}\Big|$$

where C is an absolute constant. Thus we get

$$ND_N(x,\omega) \le 1 + CU_N(x,\omega) \tag{5.8}$$

and using Lemma 5.3 we get Theorem 5.1. □

We prove now to another discrepancy result complementing Theorem 5.1.

THEOREM 5.2. *Let $X_1, X_2, \ldots$ be an i.i.d. sequence with $\mathbf{E}\,|X_1| < \infty$ and characteristic function φ. Let $S_n = X_1 + \ldots + X_n$ and put*

$$G_N(x) = 1 + \sum_{1 \le h \le \sqrt{N}} \frac{1}{h|1-\varphi(hx)|^{1/2}}.$$

Then for any fixed x

$$D_N(x,\omega) = \mathcal{O}(N^{-1/2}(\log N)^{1/4+\varepsilon} G_{2N}(x)) \qquad \textit{for a.e. } \omega.$$

PROOF. By the Erdős–Turán inequality we have

$$nD_n(x,\omega) \le C\left(\sqrt{n} + \sum_{h=1}^{[\sqrt{n}]} \frac{1}{h}|\Theta_n(h,x,\omega)|\right). \tag{5.9}$$

where

$$\Theta_n(h,x,\omega) = \sum_{\ell=1}^{n} e^{2i\pi h S_\ell(\omega) x}$$

is the quantity defined by (5.3), just making the dependence on ω explicit. Thus

$$\max_{1\le n\le 2^k} nD_n(x,\omega) \le C\Big(2^{k/2} + \sum_{h=1}^{[2^{k/2}]} \frac{1}{h} \max_{1\le n\le 2^k} \big|\Theta_n(h,x,\omega)\big|\Big). \tag{5.10}$$

By the fourth moment estimate in the first line of p. 364 of the paper of Blum and Cogburn [6] we have

$$\mathbf{E}_\omega \big|\Theta_n(h,x,\omega)\big|^4 \le C\frac{1}{|1-\varphi(hx)|^2} n^2.$$

The same moment bound holds for the translated sums $\sum_{m+1\le \ell\le m+n} e^{2\pi i h S_\ell(\omega)x}$ and thus applying Lemma 4.1 we get

$$\mathbf{E}_\omega \max_{1\le n\le 2^k} \big|\Theta_n(h,x,\omega)\big|^4 \le C\frac{1}{|1-\varphi(hx)|^2} 4^k$$

or equivalently

$$\Big\|\max_{1\le n\le 2^k} |\Theta_n(h,x,\omega)|\Big\|_4 \le C'\frac{1}{|1-\varphi(hx)|^{1/2}} 2^{k/2}.$$

Substituting this into (5.10) it follows that

$$\Big\|\max_{1\le n\le 2^k} nD_n(x,\omega)\Big\|_4 \le C''\Big(2^{k/2} + \sum_{h=1}^{[2^{k/2}]} \frac{1}{h|1-\varphi(hx)|^{1/2}} 2^{k/2}\Big) = C'' 2^{k/2} G_{2^k}(x).$$

Thus

$$\mathbf{P}_\omega\Big\{\max_{1\le n\le 2^k} nD_n(x,\omega) \ge 2^{k/2} k^{1/4+\varepsilon} G_{2^k}(x)\Big\}$$
$$\le \frac{\mathbf{E}_\omega\,(\max_{1\le n\le 4^k} nD_n(x,\omega))^4}{4^k k^{1+4\varepsilon} G_{2^k}(x)^4} = \mathcal{O}(k^{-(1+4\varepsilon)}).$$

Hence the theorem follows from the monotonicity of G_N and the Borel–Cantelli lemma. □

REMARKS. It is interesting to compare the bound obtained in Theorem 5.2 with the one obtained in Weber [50]. The two bounds are similar: the only difference is that instead of the discrepancy bound

$$\frac{1}{\sqrt{N}}\Big(\sum_{1\le h\le\sqrt{N}} \frac{1}{h|1-\varphi(hx)|}\Big)^{1/2} (\log N)^{3/2+\varepsilon} \tag{5.11}$$

obtained in [50] under minor convexity assumptions, the bound in Theorem 5.2 is

$$\frac{1}{\sqrt{N}}\Big(\sum_{1\le h\le\sqrt{2N}} \frac{1}{h|1-\varphi(hx)|^{1/2}}\Big) (\log N)^{1/4+\varepsilon}. \tag{5.12}$$

In the metric case (i.e. when we wish to estimate the discrepancy of $\{S_n(\omega)x\}$ for almost every (x,ω)), the situation simplifies considerably, and both expressions can be evaluated easily. Assume e.g. that the random variable X_1 is lattice valued and has a finite, nonzero mean c. Since scaling does not affect metric discrepancy behavior, we can assume that the span of the lattice is 2π. Then the characteristic function φ of X_1 has period 1 and it satisfies $\varphi(t) = 1 + ict + o(t)$ as $t \to 0$, and thus $|\varphi(t) - 1| \geq C|t|$ for $|t| \leq t_0$. Hence (5.12) is bounded above by

$$\frac{C}{\sqrt{N}}\Bigg(\sum_{1 \leq h \leq \sqrt{2N}} \frac{1}{h|||hx|||^{1/2}}\Bigg)(\log N)^{1/4+\varepsilon},$$

where $|||t|||$ denotes the distance of t from the nearest integer. Given a nondecreasing positive function ψ on the positive integers we say, using standard terminology in Diophantine approximation, that the irrational number x is of type $< \psi$ if $n|||nx||| \geq 1/\psi(n)$ for all positive integers n. By a well known result (see e.g. Kuipers and Niederreiter [32], p. 130, Exercise 3.5), almost every x has type $< \psi$ with $\psi(n) = \text{const} \cdot (\log n)^{1+\varepsilon}$. Also, a trivial modification of the proof of Lemma 3.3 in [32], p. 123 shows that if the type of x is $< \psi$, then

$$\sum_{h=1}^{m} \frac{1}{h|||hx|||^{1/2}} = \mathcal{O}\Bigg(\sqrt{\psi(2m)} + \sum_{h=1}^{m} \frac{\sqrt{\psi(2h)}}{h}\Bigg).$$

Using this, (5.12) yields the metric bound

$$D_N(x,\omega) = \mathcal{O}\big(N^{-1/2}(\log N)^{7/4+\varepsilon}\big) \quad \text{for almost every } (x,\omega),$$

which is better than the bound given by Theorem 5.1.

CHAPTER 6

Some open problems

In this section we mention a few open questions related to the problems investigated in Sections 1–5, whose solution would be particularly helpful in improving our understanding of this set of issues.

PROBLEM 1. In terms of the Fourier coefficients of $f \in L^2(\mathbf{T})$, find a necessary and sufficient (or at least sharp) condition for $f(kx)$ to be a convergence system. Answer the same question for the system $f(n_k x)$ for a fixed increasing sequence (n_k) of positive integers.

There are several variants and special cases of this question which are also of considerable interest.

PROBLEM 2. Find sharp conditions for the coefficient sequence (c_k) implying that $\sum_{k=1}^{\infty} c_k f(kx)$ converges a.e. for all $f \in \mathrm{Lip}_\alpha(\mathbf{T})$, $\int_{\mathbf{T}} f(t)dt = 0$. For $0 < \alpha < 1/2$ a sufficient condition is

$$\sum_{k=1}^{\infty} c_k^2 k^\gamma < \infty \qquad \text{for some } \gamma > 1 - 2\alpha$$

(cf. Gaposhkin [14] or Corollary 2.8), while the condition

$$\sum_{k=1}^{\infty} c_k^2 (\log k)^\gamma < \infty, \qquad \gamma < 1 - 2\alpha$$

is not sufficient (see Theorem 3 of Berkes [4]). Similarly, in the case $\alpha = 1/2$ a sufficient condition is

$$\sum_{k=1}^{\infty} c_k^2 (\log k)^\gamma < \infty \qquad \text{for some } \gamma > 2 \tag{6.1}$$

(see Corollary 2.9), while the condition

$$\sum_{k=1}^{\infty} c_k^2 < \infty \tag{6.2}$$

is not sufficient (see Theorem 1 in Berkes [3]).

Answer the same question for the class $BV(\mathbf{T})$, where the unknown sharp condition lies again between (6.1) and (6.2).

PROBLEM 3. Let $f \in L^2(\mathbf{T})$, $\int_{\mathbf{T}} f(t)dt = 0$. A simple sufficient condition for $f(kx)$ to be a convergence system is that the Fourier series $f \sim \sum_{k=1}^{\infty} (a_k \cos 2\pi kx + b_k \sin 2\pi kx)$ satisfies $\sum_{k=1}^{\infty} (|a_k| + |b_k|) < \infty$. (Gaposhkin [16].) Characterize the functions f for which this condition is also necessary. By Theorem 4 of Berkes [4], this is the case if

$$f \sim \sum_{k \in H} (a_k \cos 2\pi kx + b_k \sin 2\pi kx)$$

and the elements of H are coprimes.

PROBLEM 4. Find a sharp condition for the square modulus of continuity $\omega_2(\delta)$ of $f \in L^2(\mathbf{T})$ to imply that $f(n_k x)$ is a convergence system for all sequences (n_k) of positive integers satisfying the Hadamard gap condition. By Gaposhkin [15], $\int_{\mathbf{T}} f(t)dt = 0$ and

$$\omega_2(\delta, f) = \mathcal{O}\Big(\log \frac{1}{\delta}\Big)^{-\gamma}$$

are sufficient for $\gamma > 1$, but not for $\gamma = 1/2$.

PROBLEM 5. Let $W(\mathbf{T})$ (Wintner class) denote the class of functions $f \in L^2(\mathbf{T})$ such that $\int_{\mathbf{T}} f(t)dt = 0$ and the Dirichlet series $\sum_{n=1}^{\infty} a_n n^{-s}$, $\sum_{n=1}^{\infty} b_n n^{-s}$ are bounded and regular in the half-plane $\Re(s) > 0$, where

$$f \sim \sum_{k=1}^{\infty} \big(a_k \cos 2\pi kx + b_k \sin 2\pi kx\big).$$

Prove that if $f \in \mathrm{Lip}_\alpha(\mathbf{T})$, $0 < \alpha \leq 1$, $\int_{\mathbf{T}} f(t)dt = 0$, then $f(n_k x)$ is a convergence system for all sequences (n_k) of positive integers satisfying the sub-lacunarity condition

$$n_{k+1}/n_k \geq 1 + ck^{-\gamma} \qquad (k \geq k_0) \qquad 0 < \gamma < 1/2$$

if and only if $f \in W(\mathbf{T})$. The "only if" part is proved in Theorem 3.1, while a slightly weaker form of the "if" part is contained in Corollary 2.2.

Some problems on series $\sum c_k f(n_k x)$ with random n_k:

PROBLEM 6. Let $X_1 \geq 0$ be an integer valued random variable over the probability space $(\Omega, \mathcal{A}, \mathbf{P})$ such that $\mathbf{E}X_1 < \infty$ and $\mathbf{P}(X_1 = 0) < 1$. Let $S_n = X_1 + \ldots + X_n$ and let $f \in L^2(\mathbf{T})$ belong to the Wintner class $W(\mathbf{T})$. Is it true that the series $\sum_{k=1}^{\infty} c_k f(S_k(\omega)x)$ converges for almost every $(\omega, x) \in \Omega \times \mathbf{T}$ provided $\mathbf{c} \in \ell^2$? (A weaker result is proved in Theorem 4.3.)

PROBLEM 7. Find precise criteria for $\sum_{k=1}^{\infty} c_k e^{itS_k}$ to converge in $L^4(\Omega \times \mathbf{T})$ norm where $S_k = S_k(\omega)$ a nondegenerate random walk over a probability space $(\Omega, \mathcal{A}, \mathbf{P})$. A rather restrictive sufficient condition in given in Corollary 4.1.

PROBLEM 8. By a well known result of R. C. Baker [1], for any increasing sequence (n_k) of positive integers, the discrepancy $D_N(x)$ of the sequence $\{n_k x\}_{k\geq 1}$ (mod 1) is $O(N^{1/2}(\log N)^{3/2+\varepsilon})$ for any $\varepsilon > 0$ and almost every $x \in (0,1)$. Prove that this remains valid if (n_k) is a transient lattice random walk, i.e. Theorem 5.1 is valid for all $\tau > 3/2$.

Bibliography

[1] Baker, R. C., *Metric number theory and the large sieve*, J. London Math. Soc. **24** (1981), 34–40.

[2] Berkes, I., *On the asymptotic behavior of* $\sum f(n_k x)$. *Main Theorems*, Z. Wahrscheinlichkeitstheorie verw. Gebiete **34** (1976), 319–345.

[3] Berkes, I., *Critical LIL behavior of the trigonometric system*, Trans. Amer. Math. Soc. **338** (1993), 553–585.

[4] Berkes, I., *On the convergence of* $\sum c_n f(nx)$ *and the* Lip 1/2 *class*, Trans. Amer. Math. Soc. **349** (1997), 4143–4158.

[5] Billingsley, P., *Convergence of probability measures*, Wiley, New York, 1968.

[6] Blum, J. R., Cogburn R., *On ergodic sequences of measures*, Proc. Amer. Math. Soc. **51** (1975), 359–365.

[7] Bourgain, J., *Almost sure convergence and bounded entropy*, Israel J. Math. **63** (1988), 79–97.

[8] Carleson, L., *On convergence and growth of partial sums of Fourier series*, Acta Math. **116** (1966), 135–157. ndent

[9] Cassels, J. W. S., *Some metrical theorems of Diophantine approximation III*, Proc. Cambridge Phil. Soc. **46** (1950), 219–225.

[10] Erdős, P., *On trigonometric sums with gaps*, Magyar Tud. Akad. Mat. Kut. Int. Közl. **7** (1962), 37–42.

[11] Erdős, P., Koksma J. F., *On the uniform distribution modulo 1 of sequences* $(f(n, \theta))$, Proc. Konink. Nederl. Akad. Wetensch. **52** (1949), 851–854.

[12] Gál, I. S., *A theorem concerning Diophantine approximations*, Nieuw Archief voor Wiskunde **23** (1949), 13–38.

[13] Gaposhkin, V. F., *Lacunary series and independent functions* (in Russian), Uspekhi Mat. Nauk **21/6** (1966), 3–82.

[14] Gaposhkin, V. F., *On series by the system* $\varphi(nx)$ (in Russian), Math. Sbornik **69** (1966), 328–353.

[15] Gaposhkin, V. F., *A system of convergence* (in Russian), Math. Sbornik **74** (1967), 93–99.

[16] Gaposhkin, V. F., *On convergence and divergence systems* (in Russian), Mat. Zametki **4** (1968), 253–260.

[17] Gaposhkin, V. F., *On the central limit theorem for some weakly dependent sequences* (in Russian), Teor. Verojatn. Prim. **15** (1970), 666–684.

[18] Garsia, A., *Topics in almost everywhere convergence*, Lectures in Adv. Math. 4, Markham, Chicago, 1970.

[19] Ginsberg, J., Neuwirth, J., Newman, D., *Approximation by* $\{f(kx)\}$, J. Funct. Anal. **5** (1970), 194–203.

[20] Gosselin, R., Neuwirth, J., *On Paley–Wiener bases*, J. Math. Mech. **18** (1968), 871–879.

[21] Halmos, P. R., *Lectures on ergodic theory*, Publications of the Math. Society of Japan, No. 3, 1956.

[22] Harman, G., *Metric number theory*, London Math. Soc. Monographs, No. 18, Oxford University Press, 1998.

[23] Hedenmalm, H., Lindqvist, P., Seip, K., *A Hilbert space of Dirichlet series and systems of dilated functions in* $L^2(0, 1)$, Duke Math. J. **86** (1997), 1–37.

[24] Hedenmalm, H., Lindqvist, P., Seip, K., *Addendum to "A Hilbert space of Dirichlet series and systems of dilated functions in* $L^2(0, 1)$*"*, Duke Math. J. **99** (1999), 175–178.

[25] Hunt, R. A., *On the convergence of Fourier series*, Orthogonal expansions and their continuous analogues, Southern Illinois University Press, 1968, pp. 235–255.

[26] Ibragimov, I. A., *On the asymptotic distribution of values of certain sums* (in Russian), Vestnik Leningrad Univ. **15** (1960), 55–69.

[27] Kac, M., *Convergence of certain gap series*, Ann. Math. **44** (1943), 411–415.

[28] Kac, M., *On the distribution of values of sums of type* $\sum f(2^k t)$, Ann. Math. **47** (1946), 33–49.

[29] Kac, M., *Probability methods in some problems of analysis and number theory*, Bull. Amer. Math. Soc. **55** (1949), 641–665.

[30] Kesten, H., *The discrepancy of random sequences* $\{kx\}$, Acta Arith. **10** (1964), 183–213.

[31] Khinchin, A., *Ein Satz über Kettenbrüche mit arithmetischen Anwendungen*, Math. Zeitschrift **18** (1923), 289–306.

[32] Kuipers, L., Niederreiter, H., *Uniform distribution of sequences*, Wiley, New York, 1974.

[33] Landau, E., *Vorlesungen über Zahlentheorie*, Vol. 2, S. Hirzel, Leipzig, 1927.

[34] Marstrand, J. M., *On Khinchin's conjecture about strong uniform distribution*, Proc. London Math. Soc. **21** (1970), 540–556.

[35] Nikishin, E. M., *Resonance theorems and superlinear operators* (in Russian), Uspehi Mat. Nauk **25/6** (1970), 129–191.

[36] Nikishin, E. M., *On convergence systems* (in Russian), Math. Sbornik **81** (1970), 23–38.

[37] Nikishin, E. M., *Resonance theorems and function series* (in Russian), Math. Zametki **10** (1971), 583–595.

[38] Olevskii, A. M., *Fourier series with respect to general orthogonal systems*, Ergebnisse der Mathematik und ihrer Grenzgebiete, Band **86**, Springer, New York, 1975.

[39] Philipp, W., *Some metrical theorems in number theory*, Pacific J. Math. **20** (1967), 109–127.

[40] Philipp, W., *Limit theorems for lacunary series and uniform distribution mod 1*, Acta Arith. **26** (1975), 241–251.

[41] Philipp, W., *Empirical distribution functions and strong approximation theorems for dependent random variables. A problem of Baker in probabilistic number theory*, Trans. Amer. Math. Soc. **345** (1994), 707–727.

[42] Riemann, B., *Gesammelte mathematische Werke*, 2nd edition, Teubner, Leipzig, 1892.

[43] Schatte, P., On the asymptotic uniform distribution of sums reduced mod 1. Math. Nachr. **115** (1984), 275–281.

[44] Schatte, P., On a law of the iterated logarithm for sums mod 1 with application to Benford's law, Prob. Theory Rel. Fields **77** (1988), 167–178.

[45] Schatte, P., On a uniform law of the iterated logarithm for sums mod 1 and Benford's law, Lithuanian Math. J. **31** (1991), 133–142.

[46] Spitzer, F., *Principles of random walks*, second edition, Springer, New York, 1976.

[47] Toeplitz, O., *Zur Theorie der Dirichletschen Reihen*, Amer. J. Math. **60** (1938), 880–888.

[48] Weber, M., *Entropie métrique et convergence presque partout*, Coll. "Travaux en Cours" **58**, Hermann, Paris, 1998.

[49] Weber, M., *Some examples of application of the metric entropy method*, Acta Math. Hungar. **105** (2004), 39–83.

[50] Weber, M., *Discrepancy of randomly sampled sequences of reals*, Math. Nachr. **271** (2004), 105–110.

[51] Weber, M., *A theorem related to Marcinkiewicz–Salem conjecture*, Results Math. **45** (2004), 169–184.

[52] Weyl, H., *Über die Gleichverteilung von Zahlen mod Eins*, Math. Ann. **77** (1916), 313–352.

[53] Wintner, A., *On a trigonometrical series of Riemann*, Amer. J. Math. **59** (1937), 629–634.

[54] Wintner, A., *Diophantine approximations and Hilbert's space*, Amer. J. Math. **66** (1944), 564–578.

[55] Zygmund, A., *Trigonometric series I–II*, Cambridge University Press, 1959.

Editorial Information

To be published in the *Memoirs*, a paper must be correct, new, nontrivial, and significant. Further, it must be well written and of interest to a substantial number of mathematicians. Piecemeal results, such as an inconclusive step toward an unproved major theorem or a minor variation on a known result, are in general not acceptable for publication.

Papers appearing in *Memoirs* are generally at least 80 and not more than 200 published pages in length. Papers less than 80 or more than 200 published pages require the approval of the Managing Editor of the Transactions/Memoirs Editorial Board.

As of May 31, 2009, the backiog for this journal was approximately 11 volumes. This estimate is the result of dividing the number of manuscripts for this journal in the Providence office that have not yet gone to the printer on the above date by the average number of monographs per volume over the previous twelve months, reduced by the number of volumes published in four months (the time necessary for preparing a volume for the printer). (There are 6 volumes per year, each usually containing at least 4 numbers.)

A Consent to Publish and Copyright Agreement is required before a paper will be published in the *Memoirs*. After a paper is accepted for publication, the Providence office will send a Consent to Publish and Copyright Agreement to all authors of the paper. By submitting a paper to the *Memoirs*, authors certify that the results have not been submitted to nor are they under consideration for publication by another journal, conference proceedings, or similar publication.

Information for Authors

Memoirs are printed from camera copy fully prepared by the author. This means that the finished book will look exactly like the copy submitted.

Initial submission. The AMS uses Centralized Manuscript Processing for initial submissions. Authors should submit a PDF file using the Initial Manuscript Submission form found at `www.ams.org/peer-review-submission`, or send one copy of the manuscript to the following address: Centralized Manuscript Processing, MEMOIRS OF THE AMS, 201 Charles Street, Providence, RI 02904-2294 USA. If a paper copy is being forwarded to the AMS, indicate that it is for it Memoirs and include the name of the corresponding author, contact information such as email address or mailing address, and the name of an appropriate Editor to review the paper (see the list of Editors below).

The paper must contain a *descriptive title* and an *abstract* that summarizes the article in language suitable for workers in the general field (algebra, analysis, etc.). The *descriptive title* should be short, but informative; useless or vague phrases such as "some remarks about" or "concerning" should be avoided. The *abstract* should be at least one complete sentence, and at most 300 words. Included with the footnotes to the paper should be the 2000 *Mathematics Subject Classification* representing the primary and secondary subjects of the article. The classifications are accessible from `www.ams.org/msc/`. The list of classifications is also available in print starting with the 1999 annual index of *Mathematical Reviews*. The Mathematics Subject Classification footnote may be followed by a list of *key words and phrases* describing the subject matter of the article and taken from it. Journal abbreviations used in bibliographies are listed in the latest *Mathematical Reviews* annual index. The series abbreviations are also accessible from `www.ams.org/msnhtml/serials.pdf`. To help in preparing and verifying references, the AMS offers MR Lookup, a Reference Tool for Linking, at `www.ams.org/mrlookup/`.

Electronically prepared manuscripts. The AMS encourages electronically prepared manuscripts, with a strong preference for $\mathcal{AMS}$-LaTeX. To this end, the Society has prepared $\mathcal{AMS}$-LaTeX author packages for each AMS publication. Author packages include instructions for preparing electronic manuscripts, samples, and a style file that generates the particular design specifications of that publication series. Though $\mathcal{A}_{\mathcal{M}}\mathcal{S}$-LaTeX is the highly preferred format of TeX, author packages are also available in $\mathcal{A}_{\mathcal{M}}\mathcal{S}$-TeX.

Authors may retrieve an author package for *Memoirs of the AMS* from `www.ams.org/journals/memo/memoauthorpac.html` or via FTP to `ftp.ams.org` (login as `anonymous`, enter username as password, and type `cd pub/author-info`). The *AMS Author Handbook* and the *Instruction Manual* are available in PDF format from the author package link. The author package can also be obtained free of charge by sending email to `tech-support@ams.org` (Internet) or from the Publication Division, American Mathematical Society, 201 Charles St., Providence, RI 02904-2294, USA. When requesting an author package, please specify $\mathcal{A}_{\mathcal{M}}\mathcal{S}$-LaTeX or $\mathcal{A}_{\mathcal{M}}\mathcal{S}$-TeX and the publication in which your paper will appear. Please be sure to include your complete mailing address.

After acceptance. The final version of the electronic file should be sent to the Providence office (this includes any TeX source file, any graphics files, and the DVI or PostScript file) immediately after the paper has been accepted for publication.

Before sending the source file, be sure you have proofread your paper carefully. The files you send must be the EXACT files used to generate the proof copy that was accepted for publication. For all publications, authors are required to send a printed copy of their paper, which exactly matches the copy approved for publication, along with any graphics that will appear in the paper.

Accepted electronically prepared files can be submitted via the web at `www.ams.org/submit-book-journal/`, sent via FTP, or sent on CD-Rom or diskette to the Electronic Prepress Department, American Mathematical Society, 201 Charles Street, Providence, RI 02904-2294 USA. TeX source files, DVI files, and PostScript files can be transferred over the Internet by FTP to the Internet node `ftp.ams.org` (130.44.1.100). When sending a manuscript electronically via CD-Rom or diskette, please be sure to include a message identifying the paper as a Memoir.

Electronically prepared manuscripts can also be sent via email to `pub-submit@ams.org` (Internet). In order to send files via email, they must be encoded properly. (DVI files are binary and PostScript files tend to be very large.)

Electronic graphics. Comprehensive instructions on preparing graphics are available at `www.ams.org/authors/journals.html`. A few of the major requirements are given here.

Submit files for graphics as EPS (Encapsulated PostScript) files. This includes graphics originated via a graphics application as well as scanned photographs or other computer-generated images. If this is not possible, TIFF files are acceptable as long as they can be opened in Adobe Photoshop or Illustrator. No matter what method was used to produce the graphic, it is necessary to provide a paper copy to the AMS.

Authors using graphics packages for the creation of electronic art should also avoid the use of any lines thinner than 0.5 points in width. Many graphics packages allow the user to specify a "hairline" for a very thin line. Hairlines often look acceptable when proofed on a typical laser printer. However, when produced on a high-resolution laser imagesetter, hairlines become nearly invisible and will be lost entirely in the final printing process.

Screens should be set to values between 15% and 85%. Screens which fall outside of this range are too light or too dark to print correctly. Variations of screens within a graphic should be no less than 10%.

Inquiries. Any inquiries concerning a paper that has been accepted for publication should be sent to `memo-query@ams.org` or directly to the Electronic Prepress Department, American Mathematical Society, 201 Charles St., Providence, RI 02904-2294 USA.

Editors

This journal is designed particularly for long research papers, normally at least 80 pages in length, and groups of cognate papers in pure and applied mathematics. Papers intended for publication in the *Memoirs* should be addressed to one of the following editors. The AMS uses Centralized Manuscript Processing for initial submissions to AMS journals. Authors should follow instructions listed on the Initial Submission page found at www.ams.org/memo/memosubmit.html.

Algebra to ALEXANDER KLESHCHEV, Department of Mathematics, University of Oregon, Eugene, OR 97403-1222; email: ams@noether.uoregon.edu

Algebraic geometry to DAN ABRAMOVICH, Department of Mathematics, Brown University, Box 1917, Providence, RI 02912; email: amsedit@math.brown.edu

Algebraic geometry and its applications to MINA TEICHER, Emmy Noether Research Institute for Mathematics, Bar-Ilan University, Ramat-Gan 52900, Israel; email: teicher@macs.biu.ac.il

Algebraic topology to ALEJANDRO ADEM, Department of Mathematics, University of British Columbia, Room 121, 1984 Mathematics Road, Vancouver, British Columbia, Canada V6T 1Z2; email: adem@math.ubc.ca

Combinatorics to JOHN R. STEMBRIDGE, Department of Mathematics, University of Michigan, Ann Arbor, Michigan 48109-1109; email: JRS@umich.edu

Commutative and homological algebra to LUCHEZAR L. AVRAMOV, Department of Mathematics, University of Nebraska, Lincoln, NE 68588-0130; email: avramov@math.unl.edu

Complex analysis and harmonic analysis to ALEXANDER NAGEL, Department of Mathematics, University of Wisconsin, 480 Lincoln Drive, Madison, WI 53706-1313; email: nagel@math.wisc.edu

Differential geometry and global analysis to CHRIS WOODWARD, Department of Mathematics, Rutgers University, 110 Frelinghuysen Road, Piscataway, NJ 08854; email: ctw@math.rutgers.edu

Dynamical systems and ergodic theory and complex analysis to YUNPING JIANG, Department of Mathematics, CUNY Queens College and Graduate Center, 65-30 Kissena Blvd., Flushing, NY 11367; email: Yunping.Jiang@qc.cuny.edu

Functional analysis and operator algebras to DIMITRI SHLYAKHTENKO, Department of Mathematics, University of California, Los Angeles, CA 90095; email: shlyakht@math.ucla.edu

Geometric analysis to WILLIAM P. MINICOZZI II, Department of Mathematics, Johns Hopkins University, 3400 N. Charles St., Baltimore, MD 21218; email: trans@math.jhu.edu

Geometric topology to MARK FEIGHN, Math Department, Rutgers University, Newark, NJ 07102; email: feighn@andromeda.rutgers.edu

Harmonic analysis, representation theory, and Lie theory to ROBERT J. STANTON, Department of Mathematics, The Ohio State University, 231 West 18th Avenue, Columbus, OH 43210-1174; email: stanton@math.ohio-state.edu

Logic to STEFFEN LEMPP, Department of Mathematics, University of Wisconsin, 480 Lincoln Drive, Madison, Wisconsin 53706-1388; email: lempp@math.wisc.edu

Number theory to JONATHAN ROGAWSKI, Department of Mathematics, University of California, Los Angeles, CA 90095; email: jonr@math.ucla.edu

Number theory to SHANKAR SEN, Department of Mathematics, 505 Malott Hall, Cornell University, Ithaca, NY 14853; email: ss70@cornell.edu

Partial differential equations to GUSTAVO PONCE, Department of Mathematics, South Hall, Room 6607, University of California, Santa Barbara, CA 93106; email: ponce@math.ucsb.edu

Partial differential equations and dynamical systems to PETER POLACIK, School of Mathematics, University of Minnesota, Minneapolis, MN 55455; email: polacik@math.umn.edu

Probability and statistics to RICHARD BASS, Department of Mathematics, University of Connecticut, Storrs, CT 06269-3009; email: bass@math.uconn.edu

Real analysis and partial differential equations to DANIEL TATARU, Department of Mathematics, University of California, Berkeley, Berkeley, CA 94720; email: tataru@math.berkeley.edu

All other communications to the editors should be addressed to the Managing Editor, ROBERT GURALNICK, Department of Mathematics, University of Southern California, Los Angeles, CA 90089-1113; email: guralnic@math.usc.edu.

Titles in This Series

946 **Jay Jorgenson and Serge Lang,** Heat Eisenstein series on $\mathrm{SL}_n(C)$, 2009

945 **Tobias H. Jäger,** The creation of strange non-chaotic attractors in non-smooth saddle-node bifurcations, 2009

944 **Yuri Kifer,** Large deviations and adiabatic transitions for dynamical systems and Markov processes in fully coupled averaging, 2009

943 **István Berkes and Michel Weber,** On the convergence of $\sum c_k f(n_k x)$, 2009

942 **Dirk Kussin,** Noncommutative curves of genus zero: Related to finite dimensional algebras, 2009

941 **Gelu Popescu,** Unitary invariants in multivariable operator theory, 2009

940 **Gérard Iooss and Pavel I. Plotnikov,** Small divisor problem in the theory of three-dimensional water gravity waves, 2009

939 **I. D. Suprunenko,** The minimal polynomials of unipotent elements in irreducible representations of the classical groups in odd characteristic, 2009

938 **Antonino Morassi and Edi Rosset,** Uniqueness and stability in determining a rigid inclusion in an elastic body, 2009

937 **Skip Garibaldi,** Cohomological invariants: Exceptional groups and spin groups, 2009

936 **André Martinez and Vania Sordoni,** Twisted pseudodifferential calculus and application to the quantum evolution of molecules, 2009

935 **Mihai Ciucu,** The scaling limit of the correlation of holes on the triangular lattice with periodic boundary conditions, 2009

934 **Arjen Doelman, Björn Sandstede, Arnd Scheel, and Guido Schneider,** The dynamics of modulated wave trains, 2009

933 **Luchezar Stoyanov,** Scattering resonances for several small convex bodies and the Lax-Phillips conjuecture, 2009

932 **Jun Kigami,** Volume doubling measures and heat kernel estimates of self-similar sets, 2009

931 **Robert C. Dalang and Marta Sanz-Solé,** Hölder-Sobolv regularity of the solution to the stochastic wave equation in dimension three, 2009

930 **Volkmar Liebscher,** Random sets and invariants for (type II) continuous tensor product systems of Hilbert spaces, 2009

929 **Richard F. Bass, Xia Chen, and Jay Rosen,** Moderate deviations for the range of planar random walks, 2009

928 **Ulrich Bunke,** Index theory, eta forms, and Deligne cohomology, 2009

927 **N. Chernov and D. Dolgopyat,** Brownian Brownian motion-I, 2009

926 **Riccardo Benedetti and Francesco Bonsante,** Canonical wick rotations in 3-dimensional gravity, 2009

925 **Sergey Zelik and Alexander Mielke,** Multi-pulse evolution and space-time chaos in dissipative systems, 2009

924 **Pierre-Emmanuel Caprace,** "Abstract" homomorphisms of split Kac-Moody groups, 2009

923 **Michael Jöllenbeck and Volkmar Welker,** Minimal resolutions via algebraic discrete Morse theory, 2009

922 **Ph. Barbe and W. P. McCormick,** Asymptotic expansions for infinite weighted convolutions of heavy tail distributions and applications, 2009

921 **Thomas Lehmkuhl,** Compactification of the Drinfeld modular surfaces, 2009

920 **Georgia Benkart, Thomas Gregory, and Alexander Premet,** The recognition theorem for graded Lie algebras in prime characteristic, 2009

919 **Roelof W. Bruggeman and Roberto J. Miatello,** Sum formula for SL_2 over a totally real number field, 2009

918 **Jonathan Brundan and Alexander Kleshchev,** Representations of shifted Yangians and finite W-algebras, 2008

917 **Salah-Eldin A. Mohammed, Tusheng Zhang, and Huaizhong Zhao,** The stable manifold theorem for semilinear stochastic evolution equations and stochastic partial differential equations, 2008

916 **Yoshikata Kida,** The mapping class group from the viewpoint of measure equivalence theory, 2008

915 **Sergiu Aizicovici, Nikolaos S. Papageorgiou, and Vasile Staicu,** Degree theory for operators of monotone type and nonlinear elliptic equations with inequality constraints, 2008

914 **E. Shargorodsky and J. F. Toland,** Bernoulli free-boundary problems, 2008

913 **Ethan Akin, Joseph Auslander, and Eli Glasner,** The topological dynamics of Ellis actions, 2008

912 **Igor Chueshov and Irena Lasiecka,** Long-time behavior of second order evolution equations with nonlinear damping, 2008

911 **John Locker,** Eigenvalues and completeness for regular and simply irregular two-point differential operators, 2008

910 **Joel Friedman,** A proof of Alon's second eigenvalue conjecture and related problems, 2008

909 **Cameron McA. Gordon and Ying-Qing Wu,** Toroidal Dehn fillings on hyperbolic 3-manifolds, 2008

908 **J.-L. Waldspurger,** L'endoscopie tordue n'est pas si tordue, 2008

907 **Yuanhua Wang and Fei Xu,** Spinor genera in characteristic 2, 2008

906 **Raphaël S. Ponge,** Heisenberg calculus and spectral theory of hypoelliptic operators on Heisenberg manifolds, 2008

905 **Dominic Verity,** Complicial sets characterising the simplicial nerves of strict ω-categories, 2008

904 **William M. Goldman and Eugene Z. Xia,** Rank one Higgs bundles and representations of fundamental groups of Riemann surfaces, 2008

903 **Gail Letzter,** Invariant differential operators for quantum symmetric spaces, 2008

902 **Bertrand Toën and Gabriele Vezzosi,** Homotopical algebraic geometry II: Geometric stacks and applications, 2008

901 **Ron Donagi and Tony Pantev (with an appendix by Dmitry Arinkin),** Torus fibrations, gerbes, and duality, 2008

900 **Wolfgang Bertram,** Differential geometry, Lie groups and symmetric spaces over general base fields and rings, 2008

899 **Piotr Hajłasz, Tadeusz Iwaniec, Jan Malý, and Jani Onninen,** Weakly differentiable mappings between manifolds, 2008

898 **John Rognes,** Galois extensions of structured ring spectra/Stably dualizable groups, 2008

897 **Michael I. Ganzburg,** Limit theorems of polynomial approximation with exponential weights, 2008

896 **Michael Kapovich, Bernhard Leeb, and John J. Millson,** The generalized triangle inequalities in symmetric spaces and buildings with applications to algebra, 2008

895 **Steffen Roch,** Finite sections of band-dominated operators, 2008

894 **Martin Dindoš,** Hardy spaces and potential theory on C^1 domains in Riemannian manifolds, 2008